Elements of Electromigration

In this invaluable resource for graduate students and practicing professionals, Tu and Liu provide a comprehensive account of electromigration and give a practical guide on how to manage its effects in microelectronic devices, especially newer devices that make use of 3D architectures.

In the era of big data and artificial intelligence, next-generation microelectronic devices for consumers must be smaller, consume less power, cost less, and, most importantly, have higher functionality and reliability than ever before. However, with miniaturization, the average current density increases, and so does the probability of electromigration failure. This book covers all critical elements of electromigration, including basic theory, various failure modes induced by electromigration, methods to prevent failure, and equations for predicting mean-time-to-failure. Furthermore, effects such as stress, Joule heating, current crowding, and oxidation on electromigration are covered, and the new and modified mean-time-to-failure equations based on low entropy production are given. Readers will be able to apply this information to the design and application of microelectronic devices to minimize the risk of electromigration-induced failure in microelectronic devices.

This book is essential for anyone who wants to understand these critical elements and minimize their effects. It is particularly valuable for both graduate students of electrical engineering and materials science engineering and engineers working in the semiconductor and electronic packaging technology industries.

King-Ning Tu is a Professor Emeritus at UCLA and Chair Professor of Materials and Electrical Engineering at City University of Hong Kong. Professor Tu received his BSc degree from National Taiwan University, MSc degree from Brown University, and PhD degree in applied physics from Harvard University in 1960, 1964, and 1968, respectively. Professor Tu is a world leader in the science of thin films, especially in their applications in microelectronic devices, packaging, and reliability.

Dr. Yingxia Liu is an Assistant Professor at City University of Hong Kong. Dr. Liu received her PhD from the Department of Materials Science and Engineering, University of California, Los Angeles, in 2016 and her Bachelor's degree from the College of Chemistry and Molecular Engineering, Peking University in 2012.

Elements of Electromigration
Electromigration in 3D IC Technology

King-Ning Tu and Yingxia Liu

CRC Press
Taylor & Francis Group
Boca Raton London New York

CRC Press is an imprint of the
Taylor & Francis Group, an **informa** business

Designed cover image: The pancake-type void formation in a pair of solder joints caused by current crowding

First edition published 2024
by CRC Press
4 Park Square, Milton Park, Abingdon, Oxon, OX14 4RN

and by CRC Press
2385 NW Executive Center Drive, Suite 320, Boca Raton FL 33431

© 2024 King-Ning Tu and Yingxia Liu

CRC Press is an imprint of Informa UK Limited

The right of King-Ning Tu and Yingxia Liu to be identified as authors of this work has been asserted in accordance with sections 77 and 78 of the Copyright, Designs and Patents Act 1988.

British Library Cataloguing-in-Publication Data
A catalogue record for this book is available from the British Library

ISBN: 9781032470276 (hbk)
ISBN: 9781032470283 (pbk)
ISBN: 9781003384281 (ebk)

DOI: 10.1201/9781003384281

Typeset in Minion
by codeMantra

Contents

CHAPTER 1

Introduction

1.1 INTRODUCTION

In the modern era of big data and artificial intelligence, the trend of distance teaching, distance medicine, work from home, and virtual meetings (e.g., Zoom) has significantly increased the need for advanced consumer electronic products, while at the same time demanding a smaller form factor, larger memory, more function, a greater and faster rate of data transmission, a cheaper cost, and superb reliability. These consumer electronic products are operated by a set of Si chips with the size of our fingernails. Within these chips are billions of transistors, which are linked by Cu interconnect lines to form the very large-scale integration (VLSI) of microelectronic circuits. Electromigration failure in the interconnect lines has been the most important reliability issue.

The interaction between the Si chips and human consumers is via the 3D IC in advanced electronic packaging technology, which consists of Cu redistribution lines, micro solder bumps, through-Si vias, and Cu-to-Cu direct bonds. No doubt, electromigration happens in 3D IC, and it is challenging. Thus, electromigration has become an equally important reliability issue in both chip technology and advanced electronic packaging technology.

Typically, a current density of $10^5–10^6$ A/cm^2 is conducted by the Cu lines. Under such a high current density, atomic diffusion and microstructure rearrangement are enhanced due to electron scattering, leading to void formation (open) in the cathode and extrusion formation (short) in the anode of the interconnect. The circuit will fail!

At the same time, the trend of device miniaturization demands smaller and smaller components; the cross-section of interconnect lines keeps shrinking, and as the current density keeps increasing, so does the probability of circuit failure due to electromigration. This is the reason why electromigration remains the most important reliability subject that still attracts attention and study.

On the other hand, just for comparison, no electromigration occurs in an ordinary extension cord used at home or in laboratory. This is because the cord carries a very

DOI: 10.1201/9781003384281-1

low electric current density, about 10^2 A/cm^2, and also because the ambient temperature is too low for atomic diffusion to occur in the Cu wire in a cord.

Owing to the fact that Si itself is an excellent thermal conductor, most of the Joule heating in the Cu interconnects on Si chips can be conducted away, so the current density in the Cu interconnect can go up to 10^5 A/cm^2. However, in today's advanced consumer electronic products, the high density of integration of transistors and interconnects means that Joule heating is extremely large, yet heat dissipation is much more difficult. Therefore, the interaction between Joule heating and electromigration is of critical concern.

However, we must point out that the Joule heating due to electrical conduction is waste heat, and it has no direct effect on electromigration, except a certain amount of increase in temperature. This is because it is known that below the critical length of Al or Cu short stripes, there is no electromigration damage due to the effect of backstress, yet there is electrical conduction in the short stripes. Therefore, the link between Joule heating and electromigration is broken in those short stripes.

On electron scattering, we recall the free electron model of conduction of a normal metal, which assumes that the valence electrons are free to move in the electron sea in the metal, unconstrained by the perfect lattice of atoms except the scattering due to phonon vibration and structural defects such as vacancies, impurities, dislocations, and grain boundaries. The scattering is the cause of electrical resistance and Joule heating. When the current density is high, the scattering can lead to electromigration.

When an atom is out of its equilibrium position, for example, a diffusing atom in its activated state, it possesses a very large cross-section of scattering and, in turn, a very large increase in resistance. Nevertheless, when the electric current density is low, the scattering or momentum exchange between the electrons and the diffusing atoms does not enhance the displacement of the latter, so it has very little effect on atomic diffusion. However, the scattering by electrons at a high current density, above 10^4 A/cm^2, can enhance atomic diffusion along the direction of electron flow. The enhanced atomic displacement and the accumulated effect of mass transport under the scattering of a high density of electric current are called electromigration.

It is worth pointing out that in the literature on electromigration, we prefer to use the direction of flow of electrons, which is opposite to the direction of electric flow. Therefore, the directions of atomic motion and electron motion are the same in electromigration.

Furthermore, in a device having a very dense integration of circuits, such as in a 3D IC device, the effect of heat management on or heat removal from Joule heating is the most serious yield and reliability issue, and it may become the limiting factor in VLSI in the near future of microelectronic technology. There is a positive feed-back between heat and electromigration, to be discussed in Chapter 8.

Typically, a computer server is cooled in an air-conditioned room in order to maintain the device's working temperature at about 100°C, but in a hand-held mobile device, it is hard to do so.

TABLE 1.1   Comparison between Atomic Flux and Electron Flux

Atomic Flux	Electron Flux
Chemical potential: μ	Electric potential: Φ
Chemical force: $F = -\dfrac{\partial \mu}{\partial x}$	Applied voltage: V
	Electric field: $E = -\dfrac{\partial \phi}{\partial x} = -\dfrac{\partial V}{\partial x}$
Mobility: $M = \dfrac{D}{kT}$	Electron mobility: $\mu = \dfrac{e\tau}{m}$
Drift velocity: $v = MF$	Drift velocity: $v = \mu E$
Atomic flux: $J = Cv = CMF$	
Viscosity (friction coefficient): $1/M$	Electron flux: $j = nev = ne\mu E = \dfrac{E}{\rho}$ (electron current density)
Divergency:	Resistivity: $\rho = \dfrac{1}{ne\mu} = \dfrac{m}{ne^2\tau}$
$\nabla \cdot J = \dfrac{\partial J}{\partial x} + \dfrac{\partial J}{\partial y} + \dfrac{\partial J}{\partial z} = -\dfrac{\partial C}{\partial t}$	Divergency (Gauss' theorem):
	$\nabla \cdot j = -\dfrac{\partial(ne)}{\partial t}$

Electromigration involves the interaction between atomic diffusion and electron flow. We must consider both atomic flux and electron flux, as well as their coupling, due to the cross-effect in irreversible processes. It is helpful to make a direct comparison of the parameters that have been used to define these two kinds of flux. In Table 1.1, we list them side by side for comparison.

1.2 BRIEF HISTORY OF ELECTROMIGRATION IN MICROELECTRONIC TECHNOLOGY

Electromigration is the core reliability issue in microelectronic devices, which are current-voltage devices. To operate them, we must apply a high electrical current density. When the density is above 10^5 A/cm², the microstructure in the device becomes unstable due to the scattering of atoms by electrons, which can lead to an open circuit because of void formation at the cathode end of the interconnect line. The microstructure damage is defined as electromigration failure, which has been the most critical and persistent reliability problem in microelectronic devices because we cannot get rid of it and must learn to live with it [1–8].

In 1965, electromigration-induced failures in Al interconnects were found to occur in computers. It has led to many systematic studies of the topic, especially electromigration in thin-film Al interconnect lines. What was unique about electromigration in Al interconnects is that the kinetic rate is rather fast because it occurs along the grain boundaries in Al lines. This is because the operation temperature of Si devices is around 100°C, and grain boundary diffusion is dominant in Al at such a temperature. No doubt, a big effort was undertaken to find out how to slow down grain boundary diffusion in Al. It was found that the grain boundary diffusion can be greatly reduced by adding about 2 at.% of Cu to the Al. Clearly, we cannot add a higher percentage of Cu because it will increase the resistivity of the Al line.

In the late 1980s, Al interconnect technology was replaced by Cu interconnect technology. This was because when Moore's law doubles transistor density every 18–24 months, multi-layered Al interconnects, closed to ten layers, were eventually needed, which caused yield as well as Joule heating problems. Hence, Cu interconnect technology was developed to replace Al because of its lower resistance. No doubt, electromigration in Cu lines was of concern. However, it was found that electromigration in Cu occurs on the top surface of the multi-layered Cu damascene interconnect structure. This is because to produce the multi-layered interconnect microstructure of Cu, chemical-mechanical polish is used to planarize the surface before building a new layer of Cu on it. The top surface of the multi-layered Cu structure becomes the highway of electromigration because, around the operation temperature of 100°C, surface diffusion is dominant in Cu lines. Also, the deposition of a non-epitaxial cover layer of SiON on Cu does not slow down surface electromigration.

In the early 1990s, owing to the need for a large number of input/output counts for better signal resolution or voice and image resolution, flip-chip solder joints were induced to cover the entire surface of a Si chip rather than just the peripheral area of the chip. This is because the melting and solidification of solder joints do not introduce stress to the transistors below the joints. However, it was found that electromigration can occur in solder joints when the current density is near 10^4 A/cm². This is because solder is a low-melting-point alloy, so atomic diffusion in the lattice of solder is fast at the device operation temperature of 100°C, so electromigration occurs in solder joints. Therefore, electromigration has become a major reliability issue in both chip technology and electronic packaging technology.

1.3 ELECTROMIGRATION IN 3D IC TECHNOLOGY

Now, upon the end of Moore's law on miniaturization of two-dimensional integrated circuits (2D IC), three-dimensional integrated circuit (3D IC) by using chip stacking technology is the most promising way to improve Si chip performance continuously. In 3D IC, certain new microstructure elements have been introduced. First is the through-Si-vias (TSV), second is the solder micro-bump (μ-bump), third are the redistribution layers, and fourth is the Cu-to-Cu direct bonding. We need to consider electromigration failure as well as any other kinds of failure in all of them [9–12].

At the moment, the diameter of TSV is about 5–10 μm, so the current density is low; however, when the diameter is down to 1 μm, we will need to study electromigration in TSV. Furthermore, because of the large thermal expansion coefficient between Cu and Si, there will be a serious stress effect on both of them during temperature cycling. An obvious one is the pop-up of Cu TSV. Also, the stress gradient may induce creep failure in TSV.

Owing to the downward scaling of flip-chip solder joints, when the pitch is below 20 μm, the solder joint may become a Cu-Sn intermetallic compound completely, and electromigration can lead to porous joint formation. Furthermore, the downward scaling has led to another serious reliability problem because of the very large temperature gradient. Let us consider a μ-bump of 20 μm in diameter, which has a temperature difference of 2°C across the bump. The temperature gradient is 1,000°C/cm, which is very large, yet this kind of temperature gradient is rather common in 3D IC packaging technology. In other words, thermomigration becomes a critical issue in 3D IC technology and must be studied.

Hence, there exists a dilemma: if the temperature is uniform, no heat dissipation can occur and the temperature will increase. However, when there is a temperature gradient for heat dissipation, it can lead to thermomigration failure of the 3D IC packaging structure due to very large temperature gradients.

On the redistribution layer, the major concern is that the layer is open to ambient during operation. Thus, electromigration and oxidation co-exist, but their interaction has not been studied, and it is known to be a difficult problem.

On Cu-to-Cu direct bonding, the requirement is because when the μ-bump pitch is below 20 μm, reflow can lead to shorting due to direct contact of the molten solder. Hence, Cu-to-Cu direct bonding has been introduced to replace μ-bump in order to scale down the joint diameter. However, electromigration in Cu-to-Cu bonding can occur as expected.

To form Cu-to-Cu direct bonding, it is a solid-state process, no longer a solid-liquid interfacial diffusion reaction (SLID) as in flip-chip solder joint formation. Essentially, interfacial void formation is the major intrinsic reliability problem in Cu joints. Furthermore, when the Cu-to-Cu joints are stressed by a high current density, about 10^6 A/cm^2, electromigration takes place, which can lead to failure. The electromigration damage of Cu-to-Cu bonds may cause void formation at the Cu/SiO$_2$ and Cu/adhesion layer interfaces as well as at the Cu-to-Cu bonding interfaces. Void formation at the bonding interface is a new reliability issue.

What has been shown in the above is that we cannot avoid electromigration in chip technology as well as in 3D IC packaging technology, and we must learn to live with it. Hence, this book attempts to cover all the critical elements of electromigration, from the understanding of basic theory to the observation of various failure modes, as well as the method to prevent them, and the equation to predict their mean-time-to-failure.

In the ensuing chapters, the driving force and kinetic flux of electromigration will be covered, and then various failure modes induced by electromigration will be given, followed by the presentation of irreversible processes to discuss the cross-effects of microscopic reliability.

Finally, effects such as stress, Joule heating, current crowding, temperature gradient, and oxidation on electromigration will be addressed, and the new and modified mean-time-to-failure equations based on low entropy production will be given.

The very critical question on I_{max}, asked by people in the microelectronics, or to ask what is the maximum current density which can be applied to an existing device for future applications, will be analysed and answered.

1.4 COMPARISON BETWEEN ATOMIC FLUX AND ELECTRONIC FLUX

Typically, a flux is defined as the number of atoms (or charges) passing through a unit area per unit time. Thus, we have the flux, $J =$ number of atoms (or charge)/cm²-sec, and we can write

$$J = C\langle v \rangle = CMF = C\left(\frac{D}{kT}\right)\left(-\frac{d\mu}{dx}\right)$$

It is worth noting that the above equation is the most important equation in this book! For atomic flux, C is the concentration of atoms, $<v> = MF$ is the drift velocity, M is the mobility $\left(M = \dfrac{D}{kT}\right)$, F is the driving force $\left(F = -\dfrac{d\mu}{dx}\right)$, and μ is the chemical potential energy. We shall discuss the driving force for electromigration later. For charge flux, the corresponding parameters will be given in the following section.

1.4.1 On Driving Force

We have defined the "driving force" in atomic motion as the negative chemical potential energy gradient, so the chemical force acting on the diffusing atom is given as $F = -\dfrac{d\mu}{dx}$, where μ is chemical potential energy.

In electrical conduction, we have defined the driving force as eE, where $E = -\dfrac{d\Psi}{dx} = -\dfrac{dV}{dx}$ is the electric field and e is the charge. We note that E is an electric field, rather than an electric force. Electric force is given as eE. It means that if we place a charge of "e" in an electric field of E, the charge will feel a force of eE acting on it to move it in the direction of E, where E is a vector. We recall that ψ or V is defined as electric potential, which is not electric potential energy. Electric potential energy is defined as $e\psi$ or eV. The driving force is eE, or $-e\dfrac{d\Psi}{dx} = -e\dfrac{dV}{dx}$.

The unit of thermal energy is given as kT (Joule or kcal), and the unit of electric energy is given as eV. In atomic diffusion, the activation energy can be given as kcal/mole or eV/atom.

1.4.2 On Mobility

On mobility, we recall that drift velocity is given as $<v> = MF$, where M is the mobility and F is the driving force. We note that the unit of mobility of a charge and the unit of mobility of an atom should be the same. Thus, charge carrier mobility μ_e is given by

$<v>/F = <v>/eE$, where $<v>$ is the drift velocity and E is the electric field, so its unit is cm^2/eV-sec. Typically, we drop the charge and take the unit to be cm^2/V-sec. Also, charge carrier mobility $\mu_e = e\tau/m^*$, where τ is the scattering time and m^* is the electron mass.

Using Newton's law that $F = ma$, we have the unit of mass as force/acceleration, which is equal to (energy/cm)/(sec^2/cm) or eVsec^2/cm^2. Again, we obtain the unit of charge carrier mobility ($= e\tau/m^*$) to be e-sec/(eVsec^2/cm^2) = cm^2/V-sec.

On the other hand, atomic mobility is D/kT, where D is the diffusivity, which has the unit of cm^2/sec and kT is the thermal energy, so the unit of atomic mobility is cm^2/Joule-sec or cm^2/eV-sec. In comparison between charge mobility and atom mobility, the charge "e" has been dropped in carrier mobility.

1.4.3 On Flux

On atomic flux or electronic flux, we note that electric current is defined as the total number of electrons (or charges) passing through the cross-section of area A'' per unit time of the conductor. Thus, $I/A'' = j$, where j is defined as electric current density. The unit of I is amp or coulomb/sec, and the unit of j is amp/cm^2 or coulomb/cm^{2}sec. Basically, coulomb is the unit of static charge, and amp (A) is the unit of moving charge, so coulomb/sec = amp.

For comparison, in atomic diffusion, the atomic flux is defined as J = number of atoms/cm^{2}sec. In electrical conduction, j or current density is defined as j = number of charges/cm^{2}sec = coulomb/cm^{2}sec = amp/cm^2 = A/cm^2.

Finally, the interaction between a flux of moving electrons and a flux of diffusing atoms will be analysed in Chapter 3 on the kinetics of electromigration.

PROBLEMS

1.1. What is the difference between 3D IC and 2.5D IC?

1.2. What is the difference between the mobilities in atomic flux and in electron flux?

1.3. What is the difference between static charge and moving charge?

1.4. What is backstress in electromigration?

1.5. Why do we consider electromigration by grain boundary diffusion in Al interconnects but by surface diffusion on Cu interconnects?

REFERENCES

[1] H. B. Huntington and A. R. Grone, "Current-induced marker motion in gold wires," *J. Phys. Chem., Solids*, 20, 76 (1961).
[2] I. A. Blech, "Electromigration in thin aluminium films on titanium nitride," *J. Appl. Phys.*, 47, 1203 (1976).
[3] I. A. Blech and C. Herring, "Stress generation by electromigration," *Appl. Phys. Lett.*, 29, 131 (1976).

[4] P. S. Ho and T. Kwok, "Electromigration in metals," *Rep. Prog. Phys.*, 52, 301 (1989).

[5] K. N. Tu, "Electromigration in stressed thin films," *Phys. Rev. B.*, 45, 1409 (1992).

[6] R. Kirchheim, "Stress and electromigration in Al-lines of integrated circuits," *Acta Met.*, 40, 309 (1992).

[7] R. S. Sorbello, "Theory of electromigration," in *Solid State Physics*, eds by H. Ehrenreich and F. Spaepen, Academic Press, New York, Vol. 51, 159 (1997).

[8] P. C. Wang, G. S. Cargill III, I. C. Noyan, and C. K. Hu, "Electromigration induced stress in Al conductor lines measured by x-ray microdiffraction," *Appl. Phys. Lett.*, 72, 1296 (1998).

[9] T. H. Kim, M. M. R. Howlader, T. Itoh, and T. Suga, "Room temperature Cu-Cu direct bonding using surface activated bonding method," *J. Vac. Sci. Technol. A*, 21, 449–453 (2003). https://doi.org/10.1116/1.1537716.

[10] J. Y. Juang, C. L. Lu, Y. J. Li, P.N. Hsu, N. T. Tsou, K. N. Tu, and C. Chen, "A solid state process to obtain high mechanical strength in Cu-to-Cu joints by surface creep on (111)-oriented nanotwins Cu," *J. Mater. Res. Technol.*, 14, 719–730 (2021). https://doi.org/10.1016/j.jmrt.2021.06.099.

[11] K. C. Shie, A. M. Gusak, K. N. Tu, and C. Chen, "A kinetic model of copper-to-copper direct bonding under thermal compression," *J. Mater. Res. Technol.*, 15, 2332–2344 (2021). https://doi.org/10.1016/j.jmrt.2021.09.071.

[12] H. C. Liu, A. M. Gusak, K. N. Tu, and C. Chen, "Interfacial void ripening in Cu-Cu joints," *Mater. Charact.*, 181, 1–9 (2021). https://doi.org/10.1016/j.matchar.2021.111459.

CHAPTER 2

Driving Forces of Electromigration

2.1 INTRODUCTION

In the last chapter, we mentioned that in an electric field, E, if we place a charge, e, in it, the charge will feel a force of "eE" to move it in the direction of E, which is a vector. Thus, the simplest way to understand the driving force of electromigration is to place an atom having a charge of Z^*e in an electric field of E, where Z^* is assumed to be the effective charge number of the atom, so the atom will feel a force of "Z^*eE" to move it in the direction of E. We can consider "Z^*eE" to be the driving force of electromigration. The magnitude of Z^* will depend on what kind of atom it is, which will be discussed below.

The driving force and kinetics of electromigration were formally analysed by Huntington [1,2]. Figure 2.1a and b are schematic diagrams that depict the diffusion of an atom in a lattice of face-centred cubic crystal. It tends to go across the activated state and exchange positions with the neighbouring vacancy. The activated state is the state between the atom and the vacancy. Under the scattering or collision by a high current density of electrons on the diffusing atom, the latter is enhanced in its rate of exchange position with the vacancy as well as in the rate of going to a new equilibrium position.

We note that the diffusing atom has a high collision cross-section at the activated state because it is away from its equilibrium position and also because it has to push the nearest neighbouring atoms apart, so the collision cross-section is greatly increased by about a factor of 10. The force of collision to enhance the motion of the diffusing atom is defined as "electron wind force" in Huntington's model. Many studies on electromigration have followed his model [3–7].

The electron wind force of electromigration is given as $F=Z^*eE=Z^*e\rho j$, where ρ is the resistivity and j is the current density. Typically, for metallic conductors such as Al

DOI: 10.1201/9781003384281-2

9

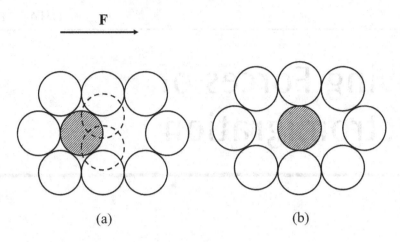

FIGURE 2.1 Schematic diagrams depicting the diffusion of an atom in a face-centred-cubic lattice.

and Cu, their value of Z^* is about one order of magnitude larger than the number of valence electrons of the normal metal atom. Therefore, Z^* for Al is about 30, and Z^* for Cu is about 10. Physically, we expect so because the cross-section of collision of a diffusing atom in the activated state is about ten times larger than that of an atom in the equilibrium state.

No doubt, the effect of electron wind force on the atomic diffusion covers the entire period of the diffusion, from the beginning of its motion to the end of exchanging positions with the vacancy, not just when the atom is in the activated state.

2.2 ELECTRON WIND FORCE OF ELECTROMIGRATION

In electromigration, the force acting on a diffusing atom is taken to be

$$F_{em} = Z^* eE = \left(Z^*_{el} + Z^*_{wd} \right) eE \tag{2.1}$$

where Z^*_{el} is the nominal valence of the diffusing atom when the dynamic screening effect is ignored, and it is responsible for the static electric field effect. $Z^*_{el}eE$ is called the direct force, and it is acting in the direction opposing the flow direction of electrons because it attracts the atom. The magnitude of Z^*_{el} is taken to be the nominal number of valence electrons in the metal atom.

Z^*_{wd} is an assumed charge number, representing the effect of momentum exchange between electrons and the diffusing atom, and $Z^*_{wd}eE$ is called the "electron wind force," and it is acting in the same direction as electron flow. Generally, Z^*_{wd} is found to be of the order of 10 of the number of valence electrons for a good conductor, so the electron wind force is much larger than the direct force for electromigration in metals. Hence, the enhanced flux of atomic diffusion is in the same direction as electron flow in electromigration. For this reason, in electromigration, we prefer to use the direction of electron flow rather than the direction of electric flow, which is opposite to the direction of electron flow.

To estimate the electron wind force, the ballistic approach to the scattering process was developed by Huntington and Grone [1]. The model postulates a transition probability of free electrons per unit time from one free electron state to another free electron state due to the scattering of the electron by the diffusing atom. The force, i.e., the momentum transfer per unit time, is calculated by summing over the initial and final states of the scattered electrons. The step-by-step derivation of the model has been presented elsewhere [1,2]. However, a simple derivation is given below.

During the elastic scattering of electrons by a diffusing atom, we assume that the system momentum is conserved. The average change in electron momentum in the transport direction equals $m_e<v>$, rather than $2m_e<v>$, where m_e is the electron mass and $<v>$ is the mean velocity of electron in the direction of electron flow. This is because the atom moves. The force on the moving atom caused by the scattering is

$$F_{wd} = \frac{m_e <v>}{\tau_{col}} \quad (2.2)$$

where τ_{col} is the mean time interval between two successive collisions.

The net momentum lost per second per unit volume of electrons to the diffusing atom is then $nm_e<v>/\tau_{col}$, and the force on a single diffusing atom is

$$F_{wd} = \frac{nm_e <v>}{\tau_{col}N_d} \quad (2.3)$$

where n is the density of electrons and N_d is the density of diffusing atoms.

The electron density can be given as $j=-ne<v>$. We recall that in atomic diffusion, we have $J=C<v>$, where J is atomic flux, C is atomic concentration, and $<v>$ is drift velocity. In comparison, we have electron flux, $j=-ne<v>$.

Substituting $<v>$ into the last equation, we obtain

$$F_{wd} = -\frac{m_e j}{e\tau_{col}N_d} = -\frac{m_e}{ne^2\tau_{col}} \frac{neE}{\rho N_d} = -\left[\frac{\rho_d}{N_d}\right]\left[\frac{n}{\rho}\right]eE \quad (2.4)$$

where $\rho=E/j$ is the resistivity of the conductor, and $\rho_d=m/ne^2\tau_{col}$ is the metal resistivity due to the diffusing atoms, and E is the applied electric field.

Besides the electron wind force, the electric field will produce a direct force on the diffusing atoms to be given by

$$F_{direct} = Z^*_{el}eE \quad (2.5)$$

where Z^*_{el} can be regarded as the nominal valence of the metal atom when the dynamical scattering effect around the atom is ignored. So the total force will be

$$F_{EM} = \left\{ Z_{el}^* - Z\left[\frac{\rho_d}{N_d}\right]\left[\frac{N}{\rho}\right] \right\} eE \tag{2.6}$$

where N is the atomic density of the conductor and $n=NZ$. The last equation can be rewritten as

$$F_{EM} = Z^* eE \tag{2.7}$$

And

$$Z^* = \left\{ Z_{el}^* - Z\left[\frac{\rho_d}{N_d}\right]\left[\frac{N}{\rho}\right] \right\} \tag{2.8}$$

where Z^* is defined as the effective charge number of the atom in electromigration.

The above model shows that in essence, the effective charge number of the atom in electromigration can be given in terms of the ratio of "specific resistivity of a diffusing atom" and "specific resistivity of a normal lattice atom":

$$Z_{wd}^* = Z\frac{\rho_d/N_d}{\rho/N} \tag{2.9}$$

where $\rho=m_0/ne^2\tau$ and $\rho_d=m^*/ne^2\tau_d$ are the resistivity of the equilibrium lattice atoms and the diffusing atoms, respectively; m_0 and m^* are the free electron mass and effective electron mass, respectively, and they are equal; and τ and τ_d are the relaxation times of a lattice atom and a diffusing atom, respectively.

In a face-centred cubic crystal lattice, there are 12 equivalent jump paths along the <110> directions. For a given electron current direction, the average resistivity of a diffusing atom must be corrected by a factor of 1/2. By rewriting the above equation, we have

$$Z^* = -Z\left[\frac{1}{2}\frac{\dfrac{\rho_d}{N_d}}{\dfrac{\rho}{N}} - 1 \right] \tag{2.10}$$

where we have taken Z_{el} as Z, the nominal valence of the metal atom. Conceptually, it means that in order to calculate Z^*, we need to know the specific resistivity ratio of a diffusing atom to a lattice atom. We shall analyse the ratio below.

2.2.1 Calculation of the Effective Charge Number

The specific resistivity of a lattice atom in a metal is assumed to be proportional to the elastic cross-section of scattering, which in turn is assumed to be proportional to the average square displacement from equilibrium, or $<x^2>$. Furthermore, the cross-section of a normal lattice atom can be estimated from the Einstein model of atomic vibration, in which the energy of each mode is

$$\frac{1}{2}m\omega^2 <x^2> = \frac{1}{2}kT \qquad (2.11)$$

where the product of $m\omega^2$ is the force constant of the vibration, and m and ω are atomic mass and angular vibrational frequency, respectively.

On the cross-section of scattering of a diffusing atom, $<x_d^2>$, we assume that the atom and its surrounding atoms, as shown in Figure 2.1b, have acquired the motion activation energy of diffusion, ΔH_m, which is independent of temperature.

$$\frac{1}{2}m\omega^2 <x_d^2> = \Delta H_m \qquad (2.12)$$

Therefore, the ratio of the last two equations gives the ratio of the cross-section of scattering,

$$\frac{<x_d^2>}{<x^2>} = \frac{2\Delta H_m}{kT} \qquad (2.13)$$

Substituting the last equation into Eq. (2.7) of Z^*, we have

$$Z^* = -Z\left[\frac{\Delta H_m}{kT} - 1\right] \qquad (2.14)$$

Now, the value of Z^* can be calculated at a given temperature.

The calculated values of Z^*, based on Eq. (2.14), agree very well with those measured for Au, Ag, Cu, Al, and Pb. Experimental measurements of Z^* will be presented in Chapter 3.

Also, at 480°C and 640°C, the measured and calculated Z^* for Al (taking $\Delta H_m = 0.62$ eV/atom) are about −30 to −26, respectively. The temperature dependence of Z^* calculated for Au is also found to agree well with the measured value by Huntington and Grone [1].

2.3 CURRENT DENSITY GRADIENT FORCE OF ELECTROMIGRATION

In a multi-layered structure of Al and Cu interconnect, electrical current needs to go up and down between layers or to turn left or right at the same level. In these cases, current crowding occurs. Often, it has been found that electromigration-induced void formation in a multi-layered structure can occur in regions with low current density, which is against what we would expect based on the electron wind force presented above.

A new driving force of electromigration, which has been called the "current density gradient force," has been introduced to explain the unexpected finding of void formation in low current density regions. This new force, or the current density gradient force, will be presented in Chapter 7.

2.3.1 Void Formation in the Low Current Density Regions

Experimental findings of void formation in low current density regions will be presented in Chapter 7 too.

PROBLEMS

2.1. What is the difference between electron wind force and current density gradient force?

2.2. Why can we present the electron wind force as $Z^* eE$ and what is the meaning of Z^*?

2.3. What is Z^* for Al, Cu and solder alloy?

2.4. Why does current density gradient force-induced void formation occur near a low current density region?

2.5. How can we define Z^* for a vacancy in Cu?

REFERENCES

[1] H. R. Huntington and A. R. Grone, "Current-induced marker motion in gold wires," *J. Phys. Chem. Solids*, 20, 76 (1961).
[2] K. N. Tu, "Solder Joint Technology," Appendix C: Derivation of Huntington's Electron Wind Force, Springer, New York (2007).
[3] P. S. Ho and T. Kwok, "Electromigration in metals," *Rep. Prog. Phys.*, 52, 301 (1989).
[4] I. A. Blech and C. Herring, "Stress generation by electromigration," *Appl. Phys. Lett.*, 29, 131 (1976).
[5] R. S. Sorbello, "Solid State Physics," Advances in Research and Applications, eds by H. Ehrenreich and F. Spaepen, Academic Press, New York, Vol. 51, 159 (1997).
[6] K. N. Tu, "Electromigration in stressed thin films," *Phys. Rev. B*, 45, 1409 (1992).
[7] K. N. Tu, C. C. Yeh, C. Y. Liu, and C. Chen, "Effect of current crowding on vacancy diffusion and void formation in electromigration," *Appl. Phys. Lett.*, 76, 988 (2000).

CHAPTER 3

Kinetics of Electromigration

3.1 INTRODUCTION

The kinetics of atomic transport under electromigration enable us to estimate the time of failure of an interconnect due to void formation or hillock formation, which can lead to an open circuit or a short circuit, respectively. This is the most important issue in microelectronic reliability, regarding the lifetime of a device in service or relating to its mean-time-to-failure (MTTF). The topic of MTTF will be presented in Chapter 7.

Typically, the operation temperature of a Si device is around 100°C, except for some special devices that may operate in a hush environment, such as under the hood of an automobile. At 100°C, we show in Table 3.1 that the dominant diffusion mechanisms in Al, Cu, and solder occur, respectively, by grain boundary diffusion, surface diffusion, and lattice diffusion. Thus, the atomic transport rate and kinetic path will be different for them as well as for their failure mode [1–3].

Below, we consider the atomic flux equation of electromigration, J_{EM}.

$$J_{EM} = C<v> = CMF = C\frac{D}{kT}Z^*eE = \frac{1}{\Omega}\frac{D}{kT}Z^*e\rho j \tag{3.1}$$

where C is the atomic concentration, $C = 1/\Omega$ for a pure metal, and Ω is the atomic volume. $<v>$ is the drift velocity and $<v> = MF$, where $M\ (= D/kT)$ is the atomic mobility and $F\ (= Z^*eE)$ is the driving force of electromigration.

TABLE 3.1   Diffusivity of Cu, Al, Pb, and Eutectic SnPb at 100°C

	Melting Point (K)	Temperature Ratio 373 K/Tm	Diffusivities at 100°C (373 K)
Cu	1,356	0.275	Surface $D_s = 10^{-12}$
Al	933	0.4	Grain boundary $D_{gb} = 6 \times 10^{-11}$
Pb	600	0.62	Lattice $D_l = 6 \times 10^{-13}$
Eutectic SnPb	456	0.82	Lattice $D_l = 2 \times 10^{-9}$ to 2×10^{-10}

DOI: 10.1201/9781003384281-3

15

Electromigration-induced failure is mainly due to void formation, which is the accumulation of vacancies. Therefore, we may consider vacancy diffusion instead of atomic diffusion. For this reason, we can convert the above equation by taking $CD = C_v D_v$, where C_v and D_v are vacancy concentration and vacancy diffusivity, respectively.

Equation (3.1) shows that if we can measure J_{EM} experimentally, we can calculate Z^* because the rest of the parameters are known. Therefore, in the early stages of studying electromigration, a large effort was spent to measure Z^*. For this reason, we introduce Blech's short-stripe test sample below.

3.2 BLECH'S SHORT-STRIPE TEST SAMPLE

Figure 3.1 depicts a schematic diagram of a short stripe of Al deposited on a long stripe of TiN that was deposited on a substrate. When an electric current is passed along the TiN, the current will detour to go in and out of the Al stripe because the latter has a lower resistivity. Upon applying a high current density, above 10^5 A/cm^2, for a short time, electromigration occurs in the Al stripe, where the cathode end shows depletion due to void formation and the anode end shows growth of a hillock, as depicted in Figure 3.1. What is unique about the short-stripe experiment is that not only can we observe directly the damages due to electromigration, but we can also measure the rate of void depletion, which will enable us to calculate the effective charge number of Z^* [4–6].

Figure 3.2 shows the SEM images of a set of Al short stripes after electromigration, where the depletion and hillock formation can be seen. Typically, it was found that the longer the stripe, the more the depletion and the larger the hillock growth. However, when the length of the Al short stripe is below a certain length or below a critical length, no electromigration damage can be observed. This has been called the short-length effect of electromigration. The effect was explained by Blech and Herring as being because of the influence of back-stress, which will be discussed in more detail later.

The Al stripes have a finite length and are typically covered by a protective Al oxide. The protective oxide does not serve as a good source and sink for vacancies; therefore, no or very little lattice shift occurs in electromigration. The topic of lattice shift will be explained in Section 3.5.

When electromigration drives more and more Al atoms to the anode end, where we may consider it to have a fixed volume, elastic compressive stress will build up there. On the other hand, at the cathode end, a tensile stress exists because more and more atoms are leaving.

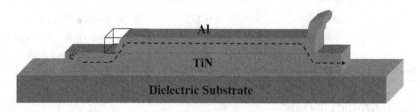

FIGURE 3.1   A schematic diagram of a short stripe of Al deposited on a long stripe of TiN which was deposited on a substrate under electromigration.

FIGURE 3.2 The morphology of Al short stripes. Left: SEM images of a set of Al short stripes after electromigration Right: SEM image of an enlarged Al stripe.

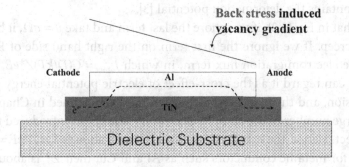

FIGURE 3.3 A schematic presentation of Blech's short stripe test sample.

On the basis of the Nabarro-Herring model of creep, there is more vacancy in the tensile region (the cathode) than the compressive region (the anode), so a vacancy concentration gradient exists in the short stripes, as depicted in Figure 3.3. The gradient will drive a flux of vacancies to go from the cathode to the anode and in turn, a flux of atoms to go from the anode to the cathode, which is against the atomic flux of electromigration. This effect is called the back-stress effect. However, because the shorter the stripe, the larger the vacancy gradient, at and below a critical length, the net effect of balancing opposing fluxes produces no electromigration damage.

Furthermore, the effect of an externally applied stress on electromigration will be the same as the back-stress effect, which will be analysed below.

3.3 INTERACTION BETWEEN STRESS AND ELECTROMIGRATION

If we consider both atomic flux and electronic flux under stress and electromigration, we have a pair of very simple equations as below:

$$J_{atom} = C\frac{D}{kT}\left(-\frac{\partial \mu}{\partial x}\right) + C\frac{D}{kT}(Z^*eE) \qquad (3.2)$$

$$J_{electron} = L_{21}X_1 + n\mu_e eE \qquad (3.3)$$

where J_{atom} is the atomic flux in units of atoms/cm²-sec, and $J_{electron}$ (= j = current density) is the electronic flux in units of coulomb/cm²-sec (which is equal to current density = current/cm²). C is the concentration of atoms per unit volume, and n is the concentration of conduction electrons per unit volume. D/kT is the atomic mobility, and μ_e is the electron mobility. The unit of D/kT is basically the same as the unit of μ_e, and they are cm²/eV-sec and cm²/V-sec, respectively, where e is the electric charge, V is the electric potential, and eV is the electric potential energy or electric energy.

In atomic flux driven by stress, we have the chemical potential energy $\mu = \sigma\Omega$, which is also the stress potential energy for atomic diffusion. In electronic flux, E is the electric field (negative gradient of electric potential) for electron transport, and $E = \rho j$, where ρ is the resistivity and j is the current density. L_{21} is the phenomenological coefficient of cross-effect due to the effect of stress potential energy gradient on electronic flux, and it contains the deformation potential [3].

We note that in Eq. (3.2), if we ignore the last term and take $\mu = \sigma\Omega$, it becomes the equation of creep. If we ignore the first term on the right-hand side of Eq. (3.2), we have only the electromigration flux term, in which $J_{atom} = C(D/kT)Z^*eE$, as given in Eq. (3.1). We can regard it as the cross-effect of electric potential energy gradient on atomic diffusion, and the parameter Z^*, which has been defined in Chapter 2 as the effective charge number of the atom in electromigration, was introduced to represent the cross-effect. Hence, the driving force of electromigration is $F = Z^*eE = Z^*e\rho j$.

Typically, for metallic conductors such as Al and Cu, their Z^* is about one order of magnitude larger than the number of valence electrons of the normal metal atom. Physically, we expect so because the cross-section of collision of a diffusing atom in the activated state is about ten times larger than that of an atom in the equilibrium state, as explained in Chapter 2.

It is worth mentioning that in order for electromigration to occur, the current density must be very high. For example, there is a critical current density, below which no electromigration will occur in Al, Cu, or solder interconnects. An example is the electric cord used at home.

We also note that in Eq. (3.3), if we ignore the L_{21} term, it becomes $\rho j = E$, which is Ohm's law, where we can take $1/\rho = ne\mu_e$. We will not discuss the L_{21} term here.

The chemical potential energy of a stressed solid is given as $\mu = \pm\sigma\Omega$, where σ and Ω are elastic normal (no shear) stress and atomic volume, respectively. In Eq. (3.2), if we let $J_{atom} = 0$, we have

$$\Delta x = \frac{\Delta\sigma\Omega}{Z^*e\rho j} \tag{3.4}$$

where Δx is defined as the critical length of short stripes below which there is no electromigration damage. It is worth mentioning that in substituting $\mu = \pm\sigma\Omega$ into Eq. (3.2), we do not consider "elastic stress flow," because there is none. Rather, we consider the effect of stress potential on equilibrium vacancy concentration and, in turn, the effect of the change in vacancy concentration on atomic flow. This is also the basic principle of elastic creep.

In Eq. (3.4), the product of "$j\Delta x$" is called the critical product, and its value is typically about 1,000 A/cm for Al interconnects. For Pb-free solder, the critical product is about 30 A/cm. However, it is worth mentioning that in solder joints, especially micro-bumps, due to the chemical reaction between solder and Cu or Ni to form intermetallic compounds (IMC), all the solder can be consumed completely to form IMC. It shows no effect of the critical length of solder, but this is because the chemical potential energy of chemical reaction is much greater than the stress potential energy. Also, it is because of the short-length effect. For this reason, we need to study electromigration in Cu-Sn intermetallic compound (IMC) joints, and porous IMC joints have been observed; see Chapter 4.

We can use Eq. (3.4) to calculate Z^* because all the other variables are measurable, and we can assume the elastic strain to be near the elastic limit. We note that the microstructure of the typical Al stripes is polycrystalline, so grain boundary diffusion dominates the electromigration at the test temperature near 200°C. Thus, the calculated Z^* should be related to grain boundary diffusion rather than lattice diffusion.

In other words, when we discuss the electromigration flux in Eq. (3.1), we need to consider not only the thermodynamic driving force but also the kinetics of diffusion. Actually, in the next chapter, when we review the experimental observations of electromigration damages in Al, Cu, and solder interconnects in microelectronic devices, we show that they are dominated by grain boundary diffusion in Al, surface diffusion on Cu, and lattice diffusion in solder at the device operation temperature of 100°C. Different kinetic paths will lead to different failure sites and modes.

3.4 ATOMIC FLUX DIVERGENCE

In electrical conduction, Kirchhoff's law states that the sum of all currents in and out of a point must be zero; thus, the in-flux equals the out-flux, so there is no electrical flux divergence at the point, except if the point is a source or sink of charges. However, atomic flux can have flux divergence at a physical point, for example, at a triple point (or a triple line) where three grain boundaries meet in a polycrystalline microstructure. If we assume the diffusivity and the effective grain boundary width of diffusion of the three are the same, there will be atomic flux divergence at the triple point where the in-flux comes from one grain boundary but the out-flux goes out along the other two grain boundaries, or vice versa. The net effect of flux divergence after a long time is the accumulation or depletion of vacancies (or atoms), which in turn leads to the formation of a void or a hillock; in other words, it becomes a circuit failure.

When the vacancy concentration is over the super-saturation concentration needed for the nucleation of a void, the ensuing growth of the void can lead to an open circuit. This has been found to occur in Al interconnects because electromigration in Al interconnects at the device operation temperature of 100°C is controlled by grain boundary diffusion. Thus, any site of atomic flux divergence due to grin boundary diffusion is of concern.

We note that if there is no atomic flux divergence, or if the atomic flux is in a steady state in an interconnect and in turn, the opposite vacancy flux is in a steady state, there will be no electromigration damage while there is atomic transport by electromigration. Furthermore, if we can assume that the vacancy concentration is at equilibrium everywhere in the sample or that the sources and sinks of vacancies are fully operational in the sample to accommodate the absorption and emission of vacancies, no void nucleation can occur because the nucleation requires a super-saturation of vacancies, and in turn, no void formation will occur.

In the classical model of interdiffusion by Darken, there is no void formation when equilibrium vacancy is assumed to exist everywhere in the sample. Below, we shall briefly discuss the flux divergence sites.

The most common site of vacancy absorption and emission is the free surface of the interconnect. However, when the surface is oxidized and if the oxide is stable and protective, as the Al oxide on Al interconnects (or Sn oxide on Sn-plated surfaces), the Al/oxide interface is no longer an effective source and sink of vacancies, so flux divergence can occur. For Cu interconnects, the free surface after chemical-mechanical planarization is the most important path of flux divergence, where electromigration damage occurs because it has no protective oxide.

The next common site of flux divergence is the interphase interface, such as the interface between a W via and an Al line. This is because solubility and diffusivity change dramatically across the interface.

Then, grain boundaries, especially the triple points, are sites of flux divergence. No doubt, we must consider dislocations, especially the kink site on an edge dislocation line. As dislocation climbs, the kink sites can absorb and emit atoms and vacancies. Consequently, lattice planes can be created or destroyed. When the created or destroyed lattice plane can migrate in the lattice as needed, it leads to lattice shift, and the event has no effect on microstructure damage or void formation. Below, we explain what a lattice shift is.

3.5 LATTICE SHIFT IN INTERDIFFUSION

Lattice shift is defined as the lattice plane migration which is manifested by marker motion in Darken's analysis of interdiffusion. Figure 3.4 is a schematic diagram to illustrate the basic idea of lattice shift in interdiffusion, where we assume there is a flux of atoms moving from the right to the left. It is accompanied by a flux of vacancies moving in the opposite direction. On the right-hand side of the figure, we assume there is a dislocation loop that serves as the sink of vacancies, so the dislocation loop will climb to increase its diameter by absorbing vacancies. Also, on the left-hand side, we assume there is a dislocation loop that serves as the source of vacancies, so the loop will shrink to decrease its diameter by emitting vacancies.

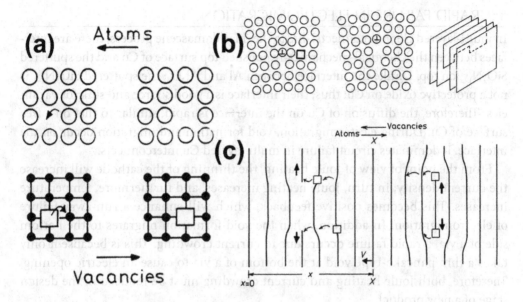

FIGURE 3.4   A schematic diagram to illustrate the idea of lattice shift in interdiffusion.

With time, these motions of dislocation will eliminate an atom plane on the right-hand side, and at the same time, they will produce an atomic plane on the left-hand side. Thus, the combined effect will cause all the atomic planes in between the two dislocation loops to shift to the right by one atomic plane thickness. This effect is defined as a "lattice shift." If we implant a marker in the middle region, it will move with the lattice shift, so we can measure the marker displacement or marker motion.

In Darken's model of the Kirkendall effect of interdiffusion, there is no void formation and no stress bending. However, in most real cases of interdiffusion and reactions, Kirkendall or Frenkal void formation occurs. Also, bending of samples due to the stress effect has been found because of insufficient lattice shift. If insufficient or no lattice shift occurs, the excess vacancies cannot be absorbed, and we have to create lattice sites to accommodate them. If they form voids, the void will take new lattice sites, thus increasing the volume. It means we will have a non-constant volume process when failure occurs due to void formation.

In the opposite direction, because we have to assume the total number of atoms is conserved, then how about the concept of lattice shift? We can create vacancies, but we cannot create atoms. Nevertheless, when we have whisker growth, we must allow new lattice sites to be created in order to allow the growth of the whisker out of the sample. Thus, the concept of failure requires flux divergence, but no lattice shift remains true.

However, if the ends of a lattice plane are pinned by a protective oxide interface, the plane cannot migrate, and flux divergence will occur. In interdiffusion, it will lead to Frenkel or Kirkendall void formation.

3.6 RAPID FAILURE IN ELECTROMIGRATION

In multi-layered Cu interconnects, due to the dual-damascene process, there are interfaces between the chemical-mechanically polished top surface of Cu and the sputtered SiO_xN_{1-x} on top. Unlike the interface between Al and Al_2O_3, the sputtered SiO_xN_{1-x} is not a protective oxide on Cu; thus, their interface is a good source and sink for vacancies. Therefore, the diffusion of Cu on the interface is rapid, similar to that on a free surface of Cu. During electromigration, void formation and migration occur on the interface. It dominates circuit failure in multi-layered Cu interconnects.

From the point of view of Joule heating, the thinning of the cathode will increase the current density; in turn, Joule heating increases, and furthermore, temperature increases. This becomes positive feedback, which will enhance a run-away failure of electromigration. In addition, when the void formation migrates to the bottom side of a via, a rapid failure occurs due to current crowding. This is because it only takes a thin pancake-type void at the bottom of a via to cause an electric opening. Therefore, both Joule heating and current crowding must be reduced in the design stage of a new product.

PROBLEMS

3.1. Why is there a critical length in Blech's short-stripe test sample? What does it mean?

3.2. What is the difference between the growth of a hillock and the growth of a whisker at the anode in electromigration?

3.3. In an Al interconnect, if we have bamboo-type grains, how does it affect electromigration?

3.4. In a Cu interconnect, if we have nano-twins, how do the twins affect electromigration in the Cu?

3.5. In a flip-chip Pb-free solder joint, if the Sn grains in the solder are oriented with their c-axis normal to the contact interface, what will be the effect on electromigration?

REFERENCES

[1] C. K. Hu and J. M. E. Harper, "Copper interconnects and reliability," *Mater. Chem. Phys.*, 52, 5 (1998).
[2] R. Rosenberg, D. C. Edelstein, C. K. Hu, and K. P. Rodbell, "Copper metallization for high performance silicon technology," *Ann. Rev. Mater. Sci.*, 30, 229 (2000).
[3] K. N. Tu, "Recent advances on electromigration in very-large-scale -integration of interconnects," *J. Appl. Phys.*, 94, 5451 (2003).
[4] S. Shingubara, T. Osaka, S. Abdeslam, H. Sakue, and T. TakaHagi, "Void formation at no current stressed area," *AIP Conf. Proc.*, 418, 159 (1998).

[5] H. Okabayashi, H. Kitamura, M. Komatsu, and H. Mori, "In-situ side-view observation of electromigration in layered Al lines by ultrahigh voltage transmission electron microscopy," *AIP Conf. Proc.*, 373, 214 (1996).

[6] K. L. Lee, C. K. Hu, and K. N. Tu, "In-situ scanning electron microscopy comparison studies on electromigration of Cu and Cu (Sn) alloys for advanced chip interconnects," *J. Appl. Phys.*, 78, 4428 (1995).

CHAPTER 4

Damage of Electromigration

4.1 INTRODUCTION

Blech's experimental observation of damage in Al short stripes is a milestone in the study of electromigration. It was followed by experimental studies of electromigration damages in Cu interconnects. Later, electromigration was found to cause damage in flip-chip solder joints as well as in 3D IC packaging systems. Recently, electromigration damages in the redistribution layer (RDL) as well as in Cu-to-Cu direct bonding have been reported [1–4,7].

Synchrotron radiation and X-ray tomography have been used to examine electromigration damage in 3D microstructures and in nanoscale microstructures [1]. Furthermore, in solder micro-bumps, because the size is small, the entire bump can be converted to intermetallic compounds (IMC) upon heating; in turn, electromigration in IMC is of concern. Indeed, it was hoped that IMC might have a much better resistance to electromigration due to its high melting point. However, porous IMC joints were observed in electromigration [5–6]. In this chapter, a systematic presentation of various kinds of damage by electromigration is presented.

In 3D IC, because of the vertical integration, new elements of the interconnect structure, such as thought-Si-via (TSV), new RDL, and μ-bumps, are introduced. Thus, the concern of electromigration damage is no longer about a single component such as a short stripe of Al interconnect or a pair of flip-chip solder joints; rather, it is about a circuit system, especially whether or not there is a weak link in the circuit system, which can cause early failures.

Figure 4.1 shows a synchrotron radiation image of a 2.5D IC test structure. The arrows show the path of applied electron, starting from a BGA solder bump as the cathode, going through the 2.5D IC structure, and coming out of another BGA solder bump as the anode [9–10]. In Figure 4.1, due to strong X-ray absorption, the top Si chip and the Si interposer below it, as well as all the polymer board and substrate, are not observable, except the metallic interconnects and solder joints. Nevertheless, the image of arrays of TSV is quite clear.

DOI: 10.1201/9781003384281-4

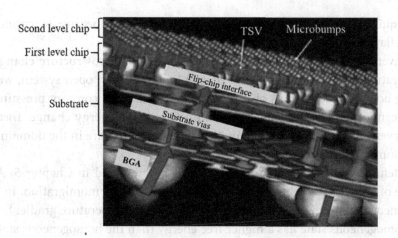

FIGURE 4.1   Synchrotron radiation image of a 2.5D IC test structure.

In electromigration tests of the 2.5D IC samples, failure is observed to occur in the new RDL, to be shown in a later section. What deserves recognition here is that the test is over an entire packaging system rather than just about a short stripe. Furthermore, what becomes important in the test is to find out the early failure that is outside the normal failure distribution as provided by the mean-time-to-failure equation. Actually, the electronic industry tends to use "burn-in" to remove early failures, which will be explained in Chapter 10. Therefore, today, in the design stage of a circuit, it must include the consideration of reliability in order to avoid any possibility of an unexpected early failure.

4.2 MICROSTRUCTURE CHANGE WITH OR WITHOUT LATTICE SHIFT

Because failure is caused by microstructure change in the device, we shall first discuss microstructure change with or without lattice shift before we present various modes of electromigration failure. The definition of lattice shift has been given in Section 3.5, Chapter 3. We note that when an irreversible process is accompanied by a lattice shift, the total number of lattice sites is conserved, it is a constant volume process, there are no voids or whisker formations, and in turn, there is no microstructure failure.

When lattice shift is absent, it is a non-constant volume process, so the excess vacancies cannot be absorbed by sinks, and extra lattice sites must be created for void formation, so microstructure failure occurs. Also, if the excess atoms cannot be absorbed, they can grow hillocks or whiskers.

We recall that phase transformations in conventional thermodynamics tend to occur in a closed system between two equilibrium states, usually under the conditions of constant temperature and constant pressure. For example, when we consider phase transformation in a eutectic SnPb solder, going from the molten state to the solid state, we can minimize Gibbs free energy change to describe the process and use the equilibrium phase diagram of SnPb to define the composition of the two solid phases in the eutectic microstructure after solidification. Nevertheless, in this special case, it is a near-equilibrium, not the final equilibrium phase change, because we cannot define the lamellar spacing in the eutectic structure.

However, when we have a phase transformation or microstructure change under a temperature gradient or a pressure (stress) gradient in an open system, we do not have the boundary conditions of constant temperature and constant pressure, so the change cannot be described by a minimum Gibbs free energy change. Instead, we have irreversible processes, so the kinetics of phase change are in the domain of non-equilibrium thermodynamics.

The definition of irreversible processes will be presented in Chapter 5. A classic example of an irreversible process is the Soret effect of thermomigration, in which a homogeneous alloy becomes inhomogeneous under a temperature gradient. Because the inhomogeneous state has a higher free energy than the homogeneous state, it is a process of increasing rather than decreasing free energy.

Nevertheless, irreversible processes may not lead to device failure if there is no atomic flux divergence. Take electromigration in an Al wire as an example; if the atomic flux driven by electromigration is uniform in the wire and if the cathode and the anode are very large sources and sinks of Al atoms, respectively, there is no failure because it is a steady state of mass transport of Al atoms from the cathode to the anode, when we define failure as void or whisker formation in the wire.

Hence, for failure to occur, we require divergence of atomic flux in irreversible processes. For example, void formation has been found at triple points of grain boundaries in Al wires, where atomic flux divergence occurs. Still, the condition of divergence is necessary but insufficient for void or whisker formation if it is a constant lattice site or constant volume process. We note that while the total number of atoms is conserved, the total number of lattice sites has to increase when there is void formation or whisker growth.

The necessary and sufficient condition of failure is that we must not have lattice shift in the region of flux divergence, so we must have a change in the number of lattice sites or a change in volume.

We recall the classical example of the Kirkendall effect in the interdiffusion between Cu and CuZn alloy. There is a divergence of flux because the flux of Cu is not equal to that of Zn, and the divergence is balanced by a flux of vacancy. Yet in Darken's analysis of interdiffusion, there is no void formation because the vacancy source and sink are assumed to be effective, so vacancies can be created or absorbed as they are needed in the microstructure. In other words, vacancies can be assumed to be at equilibrium everywhere in the sample. Thus, voids cannot nucleate if there is no supersaturation of vacancies. Furthermore, in Darken's analysis, there is no stress issue either. Hence, there is no failure while there is divergence in interdiffusion.

In a short summary, to have electromigration damage, it is required that the interconnect should be longer than the critical length so that we can overcome the back-stress effect. It must have an atomic flux divergence point where lattice shift is missing so that it is not a constant lattice site or a constant volume process. When there is interdiffusion occurring across an interface, it must not have equilibrium vacancy everywhere; otherwise, the nucleation of void cannot occur [8].

4.3 DAMAGE IN AL INTERCONNECTS

Because we have shown and discussed the damage to Al short stripes in Chapter 1, we will not repeat it here.

4.4 DAMAGE IN CU DAMASCENE INTERCONNECTS

Figure 4.2 shows the electromigration damage in Cu interconnects. It is worth noting that the Cu line is embedded in dielectric, yet we can still see the void formation on the top surface of the line. In Figure 4.2a, electrons flow from left to right, and a surface void forms on the top surface, which thins down the thickness of the Cu interconnect, which will lead to an increase in current density and in turn, a higher Joule heating rate as well as a faster electromigration rate. In Figure 4.2b, electrons go from right to left, and the void formation occurs just below the via. It will take a very short time to produce an open circuit, which is serious because of early failure.

4.5 DAMAGE IN FLIP-CHIP SOLDER JOINTS

Figure 4.3a shows a schematic diagram of the first cross-sectional view of a pair of flip-chip solder joints. Figure 4.3b shows the second cross-sectional view. The upper part of the joints is connected by an Al line, and the lower parts are connected to Cu pads. The solid arrows indicate the flow direction of electrons. In Figure 4.3c, the broken arrows indicate the flow direction of electrons in electromigration. The blue arrows indicate the direction of electromigration due to the flow of atoms, which also follows the direction of the broken arrows.

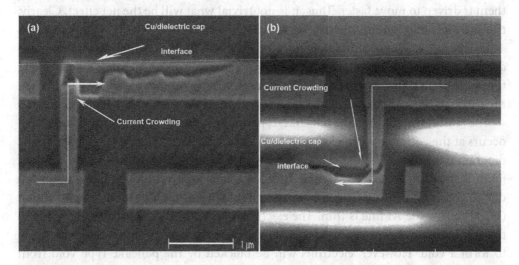

FIGURE 4.2   Surface void formation in Cu interconnects. (a) Electrons went from the left to the right. (b) Electrons went from the right to the left.

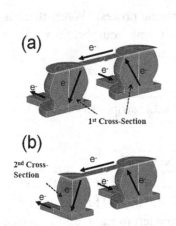

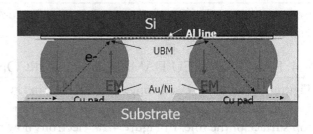

FIGURE 4.3   A schematic diagram of the cross-sectional view of a pair of flip chip solder joints. The upper part of the joints is connected by an Al line, and the lower parts are connected to Cu pads.

Owing to the fact that Joule heating in the Al line is much higher than that in the Cu pads, there is a temperature gradient across the solder joints, so the red arrows indicate thermomigration from top to bottom. We note that in the solder joint on the right-hand side, both electromigration and thermomigration are in the same direction, but in the joint on the left-hand side, they are opposite to each other. Therefore, it is expected that the rate of damage will be faster in the right-hand side solder joint. This is true if the solder joint has only a single element, such as a pure Sn joint.

However, we recall that in thermomigration, or the Soret effect of an alloy, one element goes with the temperature gradient, yet the other element goes against the temperature gradient. Furthermore, under electromigration, it is not known which of them is driven to move faster. Thus, it is not trivial what will be the net effect. Clearly, this is a topic that requires further study.

In Figure 4.4, a set of SEM images of electromigration in a chain of six flip-chip solder joints is shown. On joint nos. 1, 3, and 5, there are pancake-type voids formed along the upper-right contact area of the joint where electrons flow into the solder joint. When the pancake-type void grows across the entire contact area, an open circuit occurs.

The reason the void has the pancake-type shape is due to current crowding, which occurs at the entrance when electrons enter the solder joint. Figure 4.5 is a schematic diagram depicting the cross-sectional view of the contact of an Al line on a solder joint. It shows that electrons would enter the solder joint very much like a water fall due to current crowding. This is because the thickness of under-bump-metallization (UBM) below the Al line is thin. The electrons push atoms in the UBM as well as in the solder to move downward or to push vacancies to move upward to the entrance to form a void. However, electrons will be blocked by the pancake-type void from entering the solder joint. Therefore, electrons must move to the front end of the void in order to enter the solder joint, so the void will grow into a pancake-type shape.

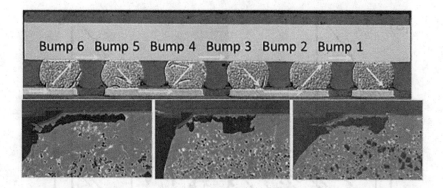

FIGURE 4.4   Pancake-type void formation in flip chip solder joints at the upper left corners of solder joints of No. 1, 3, and 5, where electrons enter into the joints.

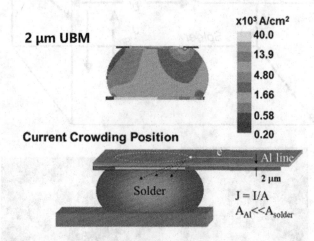

FIGURE 4.5   Simulation of current crowding in electrons going from Al line into solder joint [11].

In Figure 4.6a, a kinetic model is presented to explain the growth. In Figure 4.6b, a pair of synchrotron radiation tomographic images of a pancake-type void are shown.

To observe the microstructure change induced by electromigration, Figure 4.7a shows the cross-sectional SEM image of the formation of a pancake-type void in the upper-left corner of the solder joint. Figure 4.7b shows the SEM image when a rectangular hole is cut into the pancake-type void. Figure 4.7c shows a FIB image of the left-hand side wall of the rectangular hole, which is the second cross-sectional view of the pancake-type void, where various phases on the wall are shown. The pancake-type void is at the bottom. Below it, there is the unreacted solder. Above it, there are the IMC of Cu_6Sn_5 and Cu_3Sn, and there are a lot of Kirkendall voids in the layer of Cu_3Sn. Above the IMC are the unreacted Cu layers, and finally the Al line on top, which shows some void formation as indicated by a circle.

Figure 4.8 shows that at the opposite end of the joint, where electrons go out, a hillock or whisker grows. This is because electrons push atoms to the anode end, where the

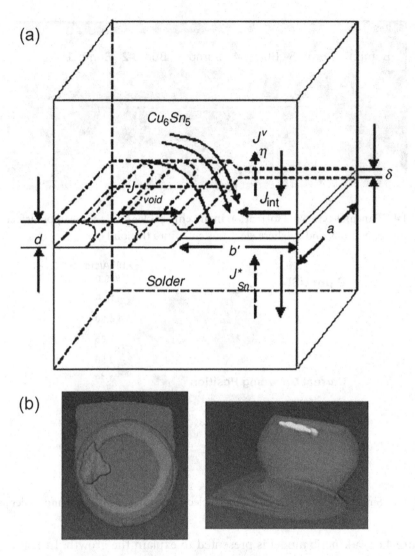

FIGURE 4.6 The formation of a pancake void (a) A kinetic model to explain the growth of a pancake-type void. (b) A pair of synchrotron x-ray tomographic images of a pancake-type of void.

compressive stress will lead to hillock growth. A detailed analysis of whisker growth will be presented in Chapter 6.

Figures 4.5 and 4.7 show the current crowding phenomenon in the contact area between an Al line and a solder joint, in which the Cu UBM is thin and may be 0.51 μm thick. However, in order to remove the current crowding phenomenon, the UBM is increased to over 10 μm. The thick Cu UBM will spread the entrance current density so that the current distribution will change from a water-fall type to a water-spring type of water drop. In turn, the pancake-type void formation will change to a distribution of small voids in the interface between the UBM and the solder. To observe the distribution of interfacial voids, we need to use synchrotron radiation tomography.

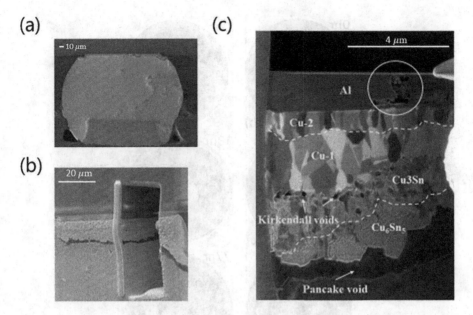

FIGURE 4.7 The morphology of a pancake void (a) The cross-sectional SEM image of the formation of a pancake-type of void in the upper-left corner of the solder joint. (b) The SEM image when a rectangular hole is cut into the pancake-type void. (c) FIB image of the left-hand side wall of the rectangular hole.

FIGURE 4.8 Electromigration induced Sn whisker growth at the upper right corner of the solder joint.

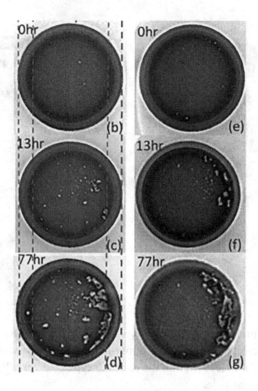

FIGURE 4.9 Two sets of interfacial images obtained by synchrotron radiation computed laminography (SRCL) for 3-dimensional examination of the early stage evolution of the size and shape of voids induced by electromigration [2].

Figure 4.9 shows two sets of interfacial images obtained by synchrotron radiation computed laminography (SRCL) for three-dimensional examination of the early stage evolution of the size and shape of a group of voids induced by electromigration. The first set was stressed at 7.5×10^3 A/cm^2 at 125°C for 0, 13, and 77 hours, respectively. The second set was stress at 1.0×10^4 A/cm^2 at 125°C for the same periods of time. The nucleation and growth of voids were analysed by the statistical model of the Weibull distribution function. Then, Johnson-Mehl-Avrami phase transformation theory was proposed to provide a physical link to the statistical damage model and to estimate the lifetime of the solder joint at the early stage of electromigration [11].

4.6 DAMAGE IN REDISTRIBUTION LAYER DUE TO JOULE HEATING

In 3D IC, because of the introduction of a Si interposer, one more layer of micro-solder bumps as well as one more RDL are added to the vertical stacking structure. When the added RDL layer is not designed properly, electromigration failure occurs.

Figure 4.10 shows a set of SEM images of electromigration-induced failure in RDL, where the failure is located in the lower part of the Si interposer. The damage is substantial, and it appears that structural melting has occurred. Indeed, when we did a simulation of Joule heating of the RDL structure, we found that there is positive feedback between electromigration and Joule heating. This is because electromigration

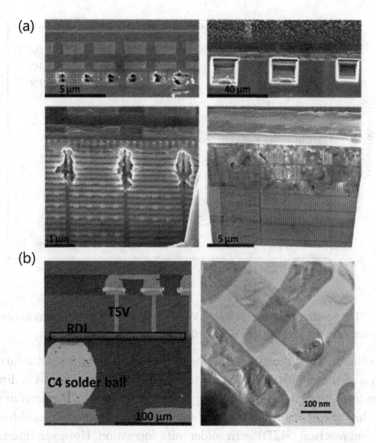

FIGURE 4.10 Morphologies of RDL. (a) A set of SEM images of electromigration induced failure in RDL which is located in the lower part of the interposer. The damage is substantial, which tells that structural melting has occurred. (b) SEM image of RDL before failure and TEM image of RDL after failure.

induces surface diffusion on the RDL, which will thin down the thickness of the RDL, which would increase the current density and increase the rate of electromigration. In turn, it would increase Joule heating to raise the temperature, which would increase the rate of electromigration as well as Joule heating again. The positive feedback can increase the temperature over the melting point of Cu, as verified by the simulation curves shown in Figure 4.11.

4.7 DAMAGE IN REDISTRIBUTION LAYER DUE TO OXIDATION AND ELECTROMIGRATION

A detailed analysis of the interaction between oxidation and electromigration will be presented in Chapter 9. Therefore, it will not be given here.

4.8 DAMAGE IN CU-TO-CU DIRECT BONDING

In 3D IC, when solder joint diameter is shrunk below 10 μm, a new reliability problem can take place. Specifically, during reflow, the molten solder may flow out and can

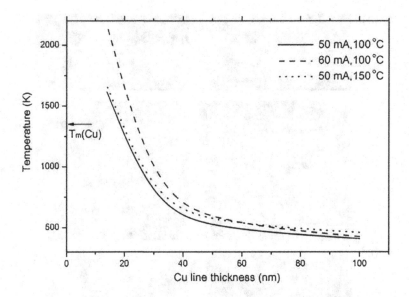

FIGURE 4.11 The positive feedback between Joule heating and electromigration can increase the temperature over the melting of Cu as verified by the simulation curves.

touch each other to cause shorts. Furthermore, the cleaning of the residue flux becomes difficult because the gap between them is too small. Therefore, Cu-to-Cu direct bonding has been introduced to replace flip-chip solder joints of small diameter or μ-bumps.

To form the Cu-to-Cu joint, it is a solid-state process, no longer a solid-liquid interfacial diffusion reaction (SLID) as in solder joint formation. However, interfacial void formation has become the major intrinsic reliability problem in Cu-to-Cu joints. The distribution function of interfacial voids and the kinetics of void evolution require further study and analysis.

On Cu-to-Cu direct bonding, low thermal budget in lowering the bonding temperature to 250°C and in reducing the bonding time to about 1 minute is a key requirement to be adopted in advanced electronic packaging for high-density interconnects (<10 μm bump diameter). Nevertheless, the bonding condition affects the bonding strength, which in turn affects the failure of electromigration.

It is well known that electromigration occurs in Cu interconnects when the current density is over 10^5 A/cm^2. This is the same for Cu-to-Cu direct bonds under electromigration. Furthermore, it has been found that bonding strength affects the outcome of electromigration. First, if the bonding strength is not strong enough to sustain the thermal mismatch between materials during electromigration tests, the bonding interface will fracture, leading to an open circuit of early failure. Second, if the bonding strength can sustain the bonding structure, voids can form at the passivation contact area between the Cu bump and the RDL due to current crowding. When the void grows along the passivation interface and separates the Cu bump and RDL, an open circuit can occur, especially when the current density and temperature are high. The bonding strength can be varied by varying the pressure and time of thermal compression bonding.

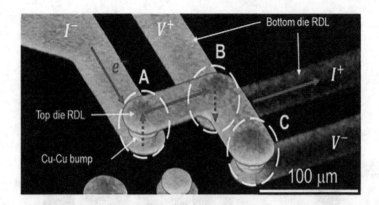

FIGURE 4.12 Kelvin structures are designed to do EM tests.

Kelvin structures are designed to do electromigration tests, as shown in Figure 4.12. It is composed of three Cu-to-Cu bumps, named as A, B, and C, and they are connected by RDLs. Current is applied and the electron flow is marked by dark arrows, so the Kelvin bumps A and B are under high current stress during the test, but only the resistance of Kelvin bump B can be measured.

The diameter of the bonding interface is 30 µm, and that of the passivation opening is 14 µm. The pitch of Cu-to-Cu bumps is 80 µm. The height of the Cu-to-Cu bump is 14 µm. The width of RDL is 15 µm, and the thickness is 3 µm. The current of the electromigration test was 1.5 A, and the chip was put on a hot plate at 150°C. The average current density at the bonding interface is 2.0×10^5 A/cm^2, and at the passivation opening is 9.8×10^5 A/cm^2. In RDL, the average current density is 1.1×10^6 A/cm^2. Thus, electromigration damages are expected to happen faster near the passivation opening and RDLs.

Because Joule heating occurred seriously during electromigration tests, the actual temperature was measured through the method of temperature coefficient of resistance (TCR). The measured temperature was slightly different from chip to chip, and the maximum value due to Joule heating was close to 50°C. The resistance of Kelvin bump B was recorded every 30 seconds during the test, and the criteria for electromigration failure is defined as a 20% resistance increase.

In Figure 4.13, the sample was bonded at 31 MPa/10 sec, and a large crack occurred along the bonding interface, so the circuit of the Kelvin bump was open. In Figure 4.14, the sample was bonded at 90 MPa/30 seconds, and a narrow void is formed under the TiW layer at the bottom of Kelvin bump B, which might have caused an open circuit in the Kelvin structure, and the resistance change was 115% after 4,035 hours of stressing time.

4.9 DAMAGE IN M-BUMP SOLDER JOINTS TO FORM POROUS IMC

In a solder micro-bump, when the solder layer is thin, it can be converted completely to Cu-Sn IMC upon annealing above 150°C. Therefore, we need to study electromigration

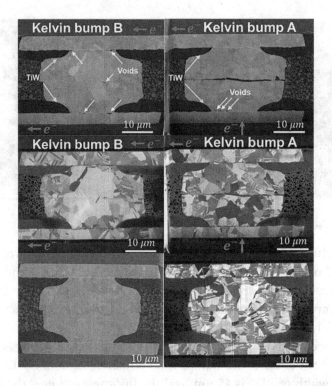

FIGURE 4.13   The sample was bonded at 31 MPa/10 sec, and a large crack has formed along the bonding interface.

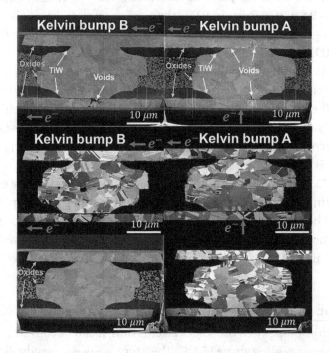

FIGURE 4.14   The sample was bonded at 90 MPa/30 sec, and a narrow void is formed under TiW layer at the bottom of Kelvin bump B [16].

in IMC in the micro-bump. What has been found, which is of wide interest, is that a porous microstructure is formed by the depletion of Sn from the Cu_6Sn_5 IMC, as shown in Figure 4.15a and b.

Electromigration tests of SnAg solder bump samples with 15 μm bump height and Cu UBM were performed. The test condition was 1.45×10^4 A/cm² at a temperature above 185°C. In the first stage, the 15 μm solder was completely converted to a sandwich layer-type structure of $Cu/Cu_3Sn/Cu_6Sn_5/Cu_3Sn/Cu$. Upon further annealing, electromigration has transformed the middle layer of Cu_6Sn_5 to a porous Cu_3Sn. The porous structure was observed to form within the bumps after several hundred hours of current stress.

In direct comparison, annealing alone at 185°C without electromigration will take more than 1,000 hours for porous Cu_3Sn to form, and it will not form at 170°C even after 2,000 hours.

A mechanism has been proposed to explain the formation of this porous structure, assisted by electromigration. The results show that the Sn was depleted from Cu_6Sn_5, which will become porous-type Cu_3Sn, and the out-diffusion of Sn will form a surface layer of Cu_3Sn on the side wall of the micro-bump. A new morphology of porous Cu_3Sn with a lamellar structure was observed to form in between the two original and layer-type Cu_3Sn, as shown in Figure 4.15b.

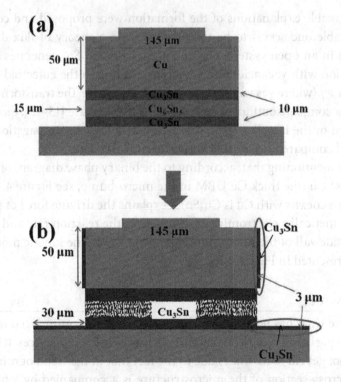

FIGURE 4.15 Schematic diagram showing the formation mechanism of the porous structure. (a) SnAg solder bump samples with 15 μm bump height and Cu under-bump-metallization (UBM) before electromigration test. (b) A porous Cu3Sn intermetallic compound (IMC) structure was observed to form within the bumps after several hundred hours of current stressing [12].

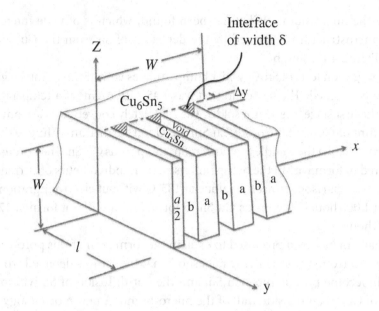

FIGURE 4.16   A schematic diagram of porous Cu3Sn intermetallic compound (IMC) structure formation.

Several possible explanations of the formation were proposed and compared. The
most reasonable one seems to be the one based on a theory of flux-driven cellular
precipitation in an open system. Outflux of Sn from Cu_6Sn_5 generates simultaneous
supersaturation with vacancies and Cu atoms, leading to the eutectoid-like transfor-
mation $\beta \rightarrow \alpha+\gamma$ (where γ is void), as shown in Figure 4.16. The transformation is complete due to a complete outflux of Sn from the Cu_6Sn_5 phase [13–15]. Simple formulae
for prediction of the lamellar structure parameters and the propagation velocity are
obtained and compared reasonably with experimental data.

It is worth mentioning that, according to the binary phase diagram of Sn-Cu, when
there is excess Cu (the thick Cu UBM in the micro-bump, see Figure 4.15), the stable
phase that can coexist with Cu is Cu_3Sn. It explains the driving force of porous Cu_3Sn
formation. Kinetically, electromigration enhances the reaction rate and drives the Sn
to go to the sidewall of the micro-bump to form Cu_3Sn. The kinetic model of the fast
reaction is presented in Figure 4.16.

4.10 SUMMARY

To summarize what has been shown above, electromigration-induced failure is com-
monly occurring in interconnect structures in microelectronic devices. It is persistence,
and we cannot get rid of it. This is due to the fact that in the trend of miniaturization,
the smaller cross-section of the microstructure is accompanied by a higher current
density, in turn resulting in a faster rate of electromigration. On the other hand, we

hope that when we understand the basic properties of electromigration, we can make our devices coexist with it.

PROBLEMS

4.1. What is a redistribution layer? Why is it needed in 3D IC?

4.2. Why must we consider a pair of solder joints rather than just one of them?

4.3. How come we do not consider thermomigration in Al and Cu interconnects?

4.4. To overcome the pancake-type void formation caused by current crowding in thin UBM, a very thick UBM is used in micro-bump technology. However, porous IMC is found. Explain the mechanism of porous IMC formation.

4.5. How come we have not found porous IMC formation in the flip-chip C4 solder joint?

REFERENCES

[1] T. Tian, K. Chen, A. A. MacDowell, D. Parkinson, Y.-C. Lai, and K. N. Tu, "Quantitative X-ray microtomography study of 3D void growth induced by electromigration in eutectic SnPb flip chip solder joints," *Script Mat.*, 65, 646–649 (2011).

[2] T. Tian, F. Xu, J. K. Han, D. Choi, Y. Cheng, L. Helfen, M. Di Michiel, T. Baumbach, and K. N. Tu, "Rapid diagnosis of electromigration induced failure time of Pb-free flip chip solder joints by high resolution synchrotron radiation laminography," *Appl. Phys. Lett.*, 99, 082114 (2011).

[3] S. Iyer, "Three-dimensional integration: an industry perspective," *MRS Bull.*, 40, 225–232 (2015).

[4] T. H. Kim, M. M. R. Howlader, T. Itoh, and T. Suga, "Room temperature Cu-Cu direct bonding using surface activated bonding method," *J. Vacuum Sci. Technol. A*, 21(2), 449–453 (2003).

[5] C. M. Liu, H. W. Lin, Y. C. Chu, C. Chen, D. R. Lyu, K. N. Chen, and K. N. Tu, "Low-temperature direct copper-to-copper bonding enabled by creep on highly (1 1 1)-oriented Cu surfaces," *Scripta Mater.*, 78, 65–68 (2014).

[6] J. Y. Juang, C. L. Lu, K. J. Chen, C. A., P. N. Hsu, C. Chen, and K. N. Tu, "Copper-to-copper direct bonding on highly (111)-oriented nanotwinned copper in no-vacuum ambient," *Sci. Rep.*, 8(1), 13910 (2018).

[7] J. R. Black, "Electromigration – a brief survey and some recent results," *IEEE Trans. Electronic Devices*, ED-16, 338–347 (1969).

[8] J. R. Lloyd, "Black's law revisited – nucleation and growth in electromigration failure," *Microelectronic Rel.*, 47, 1468–1472 (2007).

[9] K. N. Tu and A. M. Gusak, "A unified model of mean-time-to-failure for electromigration, thermomigration, and stress-migration based on entropy production," *J. Appl. Phys.*, 126, 075109 (2019).

[10] K. N. Tu and A. M. Gusak, "Mean-time-to-failure equations for electromigration, thermomigration, and stress-migration," *IEEE Trans. Comp. Packag. Manuf. Technol.*, 10, 1427–1431 (2020).

[11] W. J. Choi, E. C. C. Yeh, and K. N. Tu, "Mean-time-to-failure study of flip chip solder joints on Cu/Ni(V)/Al thin film under-bump metallization," *J. Appl. Phys.*, 94, 5665–5671 (2003).

[12] J.-A. Lin, C.-K. Lin, C.-M. Liu, Y.-S. Huang, C. Chen, D. T. Chu, and K.-N. Tu, "Formation mechanism of porous Cu3Sn intermetallic compounds by high current stressing at high temperatures in low-bump-height solder joints," *Crystals*, 6, 12 (2016).

[13] A. M. Gusak, C. Chen, and K. N. Tu, "Flux-driven cellular precipitation in open system to form porous Cu_3Sn," *Philos. Mag.*, 96, 1318–1331 (2016).

[14] C. K. Lin, C. Chen, D. T. Chu, and K. N. Tu, "Formation of porous Cu_3Sn by high-temperature current stressing," *ECS J. Solid State Sci. Technol.*, 5(9), 461–463 (2016).

[15] D. T. Chu, Y.-C. Chu, J.-A. Lin, Y.-T. Chen, C.-C. Wang, Y.-F. Song, C.-C. Chiang, C. Chen, and K.N. Tu, "Growth competition between layer-type and porous-type Cu3Sn in micro-bumps," *Microelectron. Reliab.*, 79, 32–37 (2017).

[16] K.-C. Shie, P.-N. Hsu, Y.-J. Li, K.N. Tu, C. Chen, "Effect of Bonding Strength on Electromigration Failure in Cu–Cu Bumps", *Materials*, 14, 6394 (2021). https://doi.org/10.3390/ma14216394.

CHAPTER 5

Irreversible Processes

5.1 INTRODUCTION

During the operation of electronic devices, we can have a flow of matter, a flow of energy (heat), and a flow of charged particles, or all of them together in an open system. From the viewpoint of device reliability, the most important effect is the interactions among them or their cross-effects that can lead to device failure. Especially in electromigration damage in Al and Cu interconnects, it is based on the interaction between atomic flow and electron flow. Furthermore, thermomigration in eutectic flip-chip solder joints is due to the interaction between atomic flow and heat flow, which can lead to phase separation in the eutectic microstructure of the solder alloy.

In most cases, these flows are steady. But they may not go from an equilibrium state to another equilibrium state, as in phase transformations. This is because the flow can keep going steadily in an open system. However, without equilibrium end states, we cannot use the condition of minimum Gibbs free energy to describe the process. In a steady state of flows, we are in the domain of non-equilibrium thermodynamics or irreversible processes [1–4].

In classical thermodynamics, the kinetic process going from one equilibrium state to another equilibrium state can be reversible or irreversible. Theoretically, it can be reversible. Practically, all of them tend to be irreversible.

To demonstrate the difference between a reversible and an irreversible process, Figure 5.1a depicts a closed system where a container contains water and vapour at constant temperature and constant pressure. The water and the vapour are at equilibrium, meaning that across the interface between them, there is an exchange of molecules of water and vapour, and the exchange is at micro-balance. We can say that the exchange is reversible.

Figure 5.1b depicts an open system where we open a hole in the upper piston of the container to allow vapour molecules to flow away. We also apply a constant heat source at the bottom of the container to keep the outgoing vapour leaving at a constant rate; it becomes a steady-state process, and it is irreversible.

DOI: 10.1201/9781003384281-5

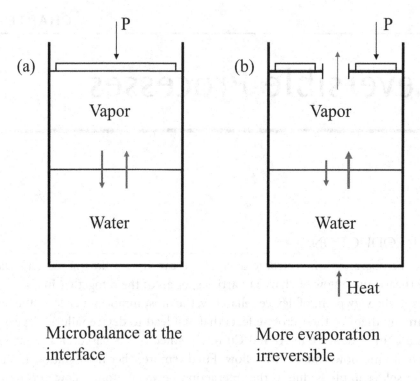

FIGURE 5.1   Schematic diagrams of reversible (a) and irreversible (b) process.

When an actual system is kept in a closed system and under homogeneous bound-ary conditions, i.e., at constant temperature and constant pressure, it goes irreversibly to an equilibrium end state, for example, in a classical phase transformation. On the other hand, if the system is kept in an open system under inhomogeneous bound-ary conditions, e.g., there is a temperature gradient or a pressure gradient, it tends to go irreversibly to a steady state instead of an equilibrium state. Typically, we call the steady-state process an "irreversible process."

5.2 FLOW IN OPEN SYSTEMS

In an irreversible process where the flow of matter, heat, or charge is in a steady state in open systems, it has been characterized by the rate of entropy change to be discussed below as an example of heat transfer. Also, we shall consider the rate of Gibbs free energy change for comparison in a later section.

It is worth noting that in a field-effect transistor, we pass an electric current or a flow of charges through the device to turn on or turn off the transistor, so it is an open system. While the total number of charges in and out of the device is conserved, entropy production is not.

In Figure 5.2, we consider two heat chambers at temperatures of T_1 and T_2, respec-tively, where the temperature T_1 is higher than T_2. If we connect them, heat will flow

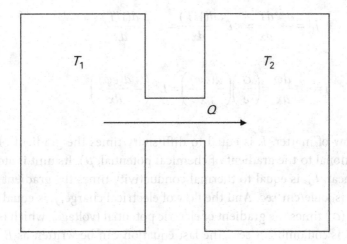

FIGURE 5.2   Schematic diagram of heat flow.

from T_1 to T_2. We assume a quantity of heat of Q flows from T_1 to T_2, and the change in entropy in the two chambers will be

$$dS_1 = -\frac{\delta Q}{T_1}$$

$$dS_2 = -\frac{\delta Q}{T_2} \tag{5.1}$$

The net change in entropy is equal to

$$dS_{net} = dS_1 + dS_2 = \delta Q\left(\frac{1}{T_2} - \frac{1}{T_1}\right) = \delta Q\left(\frac{T_1 - T_2}{TT_{21}}\right) \tag{5.2}$$

which is positive, so the heat flow has generated a certain amount of entropy. Because the heat flow takes time to go from one chamber to the other, it is a rate process of entropy generation. Below, we shall consider entropy production in a flow of matter, a flow of heat, and a flow of electrical charge. Before we do so, we need to define the flux and its driving force, as well as the cross-effects.

These three kinds of flux or flow are governed by the three well-known phenomeno-logical laws: the flow of matter (Fick's law), the flow of heat (Fourier's law), and the flow of electric charge (Ohm's law). In one dimension, they are given as follows:

$$J = -D\frac{dC}{dx} = L^{diff}\left(-\frac{d\mu}{dx}\right)$$

$$J_Q = -\kappa \frac{dT}{dx} = \kappa T^2 \frac{d(1/T)}{dx} = L^{heat} \frac{d(1/T)}{dx}$$ (5.3)

$$j = -\sigma \frac{d\varphi}{dx} = \left(\frac{\sigma}{e}\right)\left(\frac{d(e\varphi)}{dx}\right) = L^{ele}\left(-\frac{d(e\varphi)}{dx}\right)$$

where the flow of matter, J, is equal to diffusivity times the gradient of concentration (proportional to the gradient of chemical potential, μ). Its unit is atoms/cm^{2}sec. The flow of heat, J_Q, is equal to thermal conductivity times the gradient of temperature. Its unit is joule/cm^{2}sec. And the flow of electrical charge, j, is equal to electrical conductivity (σ) times the gradient of electric potential (voltage), which is the electric field. Its unit is coulomb/cm^{2}sec. The last equation can be written as $E = j\rho$, where $E = -d\phi/dx$ is electric field, ϕ is voltage, and ρ is resistivity.

The above three equations can be consolidated into a single equation of

$$J = LX$$ (5.4)

For each force X, there is a corresponding conjugate primary flow of J, and L is the proportional constant or parameter. Typically, we take the negative gradient, $-d\mu/dx$, or $-d(1/T)/dx = -(1/T^2)(dT/dx)$, or $-d(e)/dx$ as the driving force for the three flow processes, respectively.

Onsager called them conjugate forces, and the pairs of conjugate forces and the corresponding conjugate flow (or flux) will be given below. We can use a matrix to represent Eq. (5.4) in order to see the cross-effects:

$$J_i = \sum_j L_{ij} X_j$$ (5.5)

where L_{ii} are the coefficients between force and its primary flow, and L_{ij} are the coefficients of cross-effect when $i \neq j$.

In heat conduction, it has been found that a temperature gradient can also drive a flow of electric charges as well as a flow of atoms. These are defined as cross-effects. When a temperature gradient induces an electric charge flux, it is called the thermal-electrical effect, or the Seebeck effect, and the application of this effect to thermal couples for temperature measurement is well known. Reversely, when an electric field induces a heat flow, it is called the Peltier effect, which may affect interfacial reactions at the hot end of a thermal-electric device.

When a temperature gradient induces an atomic flux, it is called thermomigration, or the Soret effect. Indeed, when thermomigration occurs in a solder alloy, it is a concern for solder joint reliability. Another cross-effect that is of keen interest here is electromigration, which has been discussed in previous chapters. Its damage can lead to void formation or an open circuit at the cathode end.

When we consider heat flow, temperature is not a constant but a variable. Thus, in considering thermal-electrical effects or thermomigration, the temperature should be treated as a variable. However, atomic flow and charge carrier flow can occur at a constant temperature, so when we consider the cross-effect between them as in electromigration, we can assume a constant temperature process. Also, we may ignore the Joule heating effect when it is small. On the other hand, when Joule heating is significant, we must address it, which will be discussed in Chapter 8.

5.3 RATE OF ENTROPY PRODUCTION

To consider entropy production in a flow process, we shall deal with inhomogeneous systems. This is because there will be no flow in a homogeneous system. The main thermodynamic supposition for an inhomogeneous system is the postulation of quasi-equilibrium for physically small volumes in the system. Each physically small volume or cell can be considered to be in quasi-equilibrium, and its entropy can be determined from the classical relations of equilibrium thermodynamics by using thermodynamic variables and parameters.

According to the first law of thermodynamics, we can express a change in the internal energy of a binary system at constant temperature and constant pressure to be

$$dU = TdS - pdV + \sum_{i=1}^{2} \mu_i dn_1 \tag{5.6}$$

We note that when positive work is done by applying pressure to the system, the volume decreases, so we have the $- pdV$ term with the negative sign. Then

$$TdS = dU + pdV - \sum_{i=1}^{2} \mu_i dn_i = dH - Vdp - \sum_{i=1}^{2} \mu_i dn_i \tag{5.7}$$

Here dS, dU, dH, dV and dn_i are, respectively, the changes of entropy, internal energy, enthalpy, volume, and number of particles of i-species in a fixed physically small cell. The whole system's entropy (and its change) is defined as the sum of entropy (and their changes) of all cells.

The change in entropy in each cell can be represented as a sum of two terms: $dS = dS_e + dS_i$. The first term, dS_e, is due to the divergence of entropy flux (describing the redistribution of already available entropy between cells). It corresponds to the net difference between the entropy coming to a cell through its boundaries and the entropy going out of the cell through its boundaries. The second term, dS_i, means entropy production within the cell, and it is always positive (or zero when the system is in equilibrium and at zero fluxes).

In a steady-state process, the sum "dS" is zero in every cell, but dS_e or dS_i may not be zero. This is because a steady state means that the parameters of each cell are constant over time. Thus, entropy, which is a state function, will remain constant and dS will be zero. In a steady-state process, when the entropy production, dS_i, is not zero, it is compensated by the divergence of entropy flux, dS_e, in the cell. For example, in a steady-state process, when heat is generated in a cell, the heat must be transported away or dissipated; otherwise, the temperature will increase and it cannot be kept in a steady state.

Onsager showed that in Eq. (5.8) below, the first term is the divergence term and the second term is the rate of entropy production term. The second term can always be represented by a product of flux "J_i" and its conjugate force "X_i." The derivation of Onsager's equation for electrical conduction, atomic diffusion, and heat conduction will be given later.

$$\frac{\partial S}{\partial t} = -div(J_S) + \frac{1}{T}\sum_i J_i X_i \tag{5.8}$$

We make below a comparison between Eq. (5.8) and the atomic diffusion equation, or Fick's second law, which is

$$\frac{\partial C}{\partial t} = -\nabla \cdot J$$

The second law is based on the conservation of mass. On the other hand, if there is a source or sink of atoms in the system, we can add a source/sink term, which is optional and rare, but the last term in Eq. (5.8) on entropy production is built-in.

Below, we consider entropy production in electrical conduction, atomic diffusion, and heat conduction. Specifically, in electrical conduction, we show that entropy production is Joule heating.

5.3.1 Electrical Conduction

We consider the one-dimensional electrical conduction of a pure metal at a constant temperature and constant volume. Therefore, we have $dn_i = 0$, and $dV = 0$ in Eq. (5.7), so

$$TdS = dU \tag{5.9}$$

To consider the internal energy change in electrical conduction, dU, we start with knowing the relation in physical unit conversion that "$1\,newton \cdot meter = 1\,joule = 1\,coulomb \cdot volt$," where the unit of "newton·meter" is the unit of mechanical work, and "joule" is the unit of thermal energy, and "coulomb·volt" is the unit of electrical energy. Thus, the unit of electrical energy is eV, which is coulomb·volt. It means that the internal electrical energy change can be expressed as a product of charge and its electrical potential. Here, we define; j = current density = $\dfrac{ampere}{cm^2}$ = $\dfrac{coulomb}{cm^2}$·sec, also we define V = voltage.

Figure 5.3 depicts electric conduction in a conductor with a constant cross-section of A between x and $x + dx$, and the voltage drop is $\Delta\phi$. Thus, $jAdt$ gives the number of charge or coulomb, where dt is the time for charge to go from x to $x + dx$. If we consider a conduction of constant current density (j = constant) across a volume of $V = Adx$, which has caused a voltage drop of $\Delta\phi$, the internal energy change is given below:

$$TdS = dU = jAdt\,\Delta\phi = jAdt\left[\phi(x) - \phi(x + \Delta x)\right]$$

$$= -jVdt\left[\frac{\varphi(x + \Delta x) - \varphi(x)}{\Delta x}\right] = jVdt\left[-\frac{d\varphi}{dx}\right]$$

where $V = Adx$. Thus, we obtain

$$\frac{TdS}{Vdt} = j\left[-\frac{d\varphi}{dx}\right] = jE = j^2\rho \tag{5.10}$$

where $E = -d\phi/dx = j$ is the electric field and ρ is the resistivity. This is Onsager's equation of rate of entropy production in electrical conduction.

Therefore, in this simple case, the rate of entropy production is a product of jE, where j is the conjugate electrical current density or flux (coulomb/cm²·sec) and E is the conjugate driving force (which is the negative gradient of electric potential or the electric field). Also, j^2 is defined as "Joule heating" per unit volume per unit time. Therefore, in electrical conduction, entropy production is Joule heating. Its unit is energy/cm³·sec.

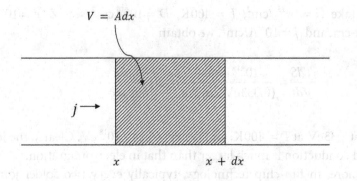

FIGURE 5.3   Electric conduction in a conductor with constant cross-sectional area.

We note that in Eq. (5.10), Joule heating alone cannot be in a steady state. While the applied current can be steady or the rate of charges being transported is constant, the entropy production increases, so the temperature without heat dissipation will go up. To reach a steady state, the system needs an outgoing heat flux of $J_Q = j^2$, which is simultaneously accompanied by the outgoing entropy fluxes of J_Q/T.

5.3.1.1 Joule Heating

In the above, we have shown that in electric conduction, entropy production is Joule heat. Usually, the power of Joule heating is given as

$$P = IV_d = I^2 R = j^2 V \tag{5.11}$$

where V_d ($= IR$) is the voltage drop, I is the applied current, and $I/A = j$, and A is the cross-sectional area of the sample, R is the resistance of the sample and $R = A/l$, and l is the length of the sample, so the volume $V = Al$. Therefore, the power of Joule heating, $I^2 R$, is Joule heating per unit time (power = energy/time) of the entire sample, and j^2 is Joule heating per unit volume per unit time of the sample.

Thus, a low power device means a low entropy production device or a device with low waste heat production.

Because Si devices operate electrically, Joule heating is a built-in cause of reliability. Below, we calculate the Joule heating due to electrical conduction as well as electromigration in an Al interconnect for comparison. If we take $j = 10^6$ A/cm^2 and $= 10^{-6} \cdot$cm for Al, the Joule heating in electrical conduction is

$$\frac{TdS}{Vdt} = jE = \rho j^2 = 10^6 \frac{\text{joule}}{\text{cm}^3\text{sec}}$$

For electromigration in an Al interconnect, we have

$$\frac{TdS}{Vdt} = JX = \left(C\frac{D}{kT}F \right)F = C\frac{D}{kT}(Z^*e\rho j)^2$$

When we take $C = 10^{23}$ /cm^3, $T = 400$K, $D = 10^{-16}$ cm^2/sec, $Z^*e = 10^{-18}$ coulomb, $\rho = 10^{-6}$ $\Omega \cdot$cm, and $j = 10^6$ A/cm^2, we obtain

$$\frac{TdS}{Vdt} = \frac{10^{-19}\text{joule}^2}{(0.033\text{eV})\text{cm}^3\text{sec}} = 20\frac{\text{joule}}{\text{cm}^3\text{sec}}$$

where $kT = 0.033$eV at $T = 400$K, and 1 joule $= 6.24 \times 10^{18}$ eV. Clearly, the Joule heating in electrical conduction is much larger than that in electromigration.

Furthermore, in flip-chip technology, typically every two solder joints are connected by an Al interconnect on the chip side, as shown in Figure 5.4a. The current density in the Al is about 10^6 A/cm^2, and the current density in the solder joint is about

(a)

(b)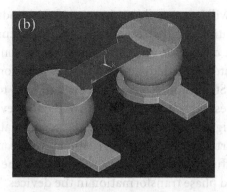

FIGURE 5.4 Simulation of current density in solder joints during the electromigration test. (a) Mesh and model of a pair of solder joints. (b) The simulation result of current density distribution in a pair of solder joints.

10^4 A/cm^2. This is because the cross-section of the Al interconnect line is about two orders of magnitude smaller than that of the solder joint. In Al conduction, we obtain Joule heating of $\rho j^2 = 10^6$ joule/cm$^3 \cdot$ sec, as shown before. For the solder joint conduction, if we take $j = 10^4$ A/cm^2 and $\rho = 10^{-5}$ $\Omega \cdot$cm for SnAgCu solder, we obtain Joule heating of $\rho j^2 = 10^3$ joule/cm$^3 \cdot$ sec, which is much less than that of Joule heating in Al. In Figure 5.4b, the simulation of current density distribution is shown.

On the basis of the above calculations, the Al interconnect is much hotter than the solder joint, and in turn, the chip side will be hotter than the substrate side. Consequently, there is a temperature gradient across the solder joint. If the temperature gradient is large enough, thermomigration can occur.

Indeed, if we consider just a temperature difference of 10 °C across a solder joint of 100 μm in diameter, the temperature gradient is 1000 °C/cm, which is very large. Therefore, temperature gradient can be huge in the microstructure of 3D IC packaging technology. Even if a temperature difference of 1 °C occurs across a solder joint of 20 μm in diameter, the temperature gradient will be 500°C/cm. In Table 5.1, we have listed the temperature gradients in a solder joint of 10 $\mu =$ μm and 20 $\mu =$ μm in thickness, and with a temperature difference of 1°C, 5°C and 10°C across the solder joint. All the calculated temperature gradients shown in Table 5.1 are large enough to cause thermomigration. Therefore, in 3D IC technology, failures caused by thermomigration must be considered.

Because Joule heating is waste heat, it cannot be used to do work; however, it generates heat and the heat can increase the temperature of the conductor. Consequently,

TABLE 5.1 Calculation of Temperature Gradient, $\Delta T / \Delta x$

$\dfrac{\Delta T}{\Delta X}$	10°C	5°C	1°C
10 μm	10^4 °C/cm	5×10^3 °C/cm	10^3 °C/cm
20 μm	5×10^3 °C/cm	2.5×10^3°C/cm	5×10^2 °C/cm

it will increase the resistance of the conductor, which in turn will cause more Joule heating. An increase in the conductor temperature will lead to thermal expansion of the conductor, and this is the basic reason for thermal stress in devices when various materials with different thermal expansion coefficients are integrated.

Stress and stress gradients can lead to fractures and creep. When the thermal stress is cyclic because semiconductor devices are being turned on and off frequently, fatigue can happen due to the accumulation of plastic strain energy in the system. Furthermore, the higher temperature caused by Joule heating will increase atomic diffusion, which in turn will increase the rate of interdiffusion, interfacial reaction, and phase transformation in the devices. Therefore, Joule heating is a major concern for device reliability, especially in 3D IC packaging technology where heat dissipation is difficult.

It is worthwhile to mention the application of Joule heating in fuses to prevent overheating in household appliances. Thin solder stripes have been used as fuses. When the appliance draws a current close to 10 Amp, the fuse will melt and the household circuit will open. If we assume the fuse has a cross-section of $2\,mm \times 0.1\,mm$, the current density in the fuse will be about 5×10^4 A/cm^2. It can melt the solder fuse. Indeed, when such a high current density was applied to solder µ-bumps in a Si device, melting of the solder bump was observed. Knowing the heat capacity of the solder, we can calculate the temperature increase due to Joule heating, provided that we know the rate of heat dissipation.

Because Si chip itself is a very good heat conductor, this is the reason why Al and Cu interconnects on Si can take a current density close to 10^6 A/cm^2, which is much higher than that in a household wire or an extension cord.

5.3.2 Atomic Diffusion

We consider a binary system in an isobaric condition,

$$TdS = dH - \sum_{i=1}^{2} \mu_i dn_i \tag{5.12}$$

Then, we take the binary system to be an element and its isotope, or an ideal solution, and then $dH = 0$. We have

$$\frac{TdS}{dt} = -\sum_{i=1}^{2} \mu_i \left(\frac{\partial n_i}{\partial t} \right)$$

If we divide the above equation by volume "V" and take $n_i/V = C_i$ and for simplicity we recall the continuity equation in one dimension, we have

$$\frac{\partial C_i}{\partial t} = -\nabla \cdot J_i = -\frac{\partial J_i}{\partial x}$$

We use the well-known differentiation of a product of $ydx = d(xy) - xdy$, and we have

$$\frac{TdS}{Vdt} = \sum_{i=1}^{2} \mu_i \frac{\partial J_i}{dx} = -\frac{\partial}{\partial x}\left(-\sum_{i=1}^{2} \mu_i J_i\right) + \sum_{i=1}^{2} J_i\left(-\frac{\partial \mu_i}{\partial x}\right)$$

(5.13)

In a steady-state process, the sum of the two terms on the right-hand side of the last equation will be zero because they compensate each other. As shown in Eq. (5.8), the first term on the right-hand side of Eq. (5.13) is the divergence term, and the second term is the rate of entropy production, which is equal to the product of flux and driving force.

5.3.3 Heat Conduction

We consider one-dimensional heat flux in a pure metal caused by a temperature gradient along the x-axis. Let the cell be a thin layer having a cross-sectional area A and a width of dx between x and $x + dx$, shown in Figure 5.5. The volume of the cell is $V = Adx$. Assuming the process to be isobaric with no atomic flow, the change in entropy in the cell is determined by the change in enthalpy, and in turn, this change of enthalpy will be determined by the difference between incoming and outgoing heat fluxes. From Eq. (5.7), we have

$$TdS = dH = J_Q(x)\cdot A\cdot dt - J_Q(x+\Delta x)\cdot A\cdot dt$$

$$= -V\frac{J_Q(x+\Delta x)-J_Q(x)}{\Delta x}\cdot dt = -V\frac{\partial J_Q}{\partial x}dt$$

(5.14)

Then we divide both sides by $TVdt$ and again use the well-known differentiation of a product; $\left[ydx = d(yx) - xdy\right]$, we have

$$\frac{\partial S}{Vdt} = -\frac{1}{T}\frac{\partial J_Q}{\partial x} = -\frac{\partial}{\partial x}\left(\frac{J_Q}{T}\right) + J_Q\frac{\partial}{\partial x}\left(\frac{1}{T}\right)$$

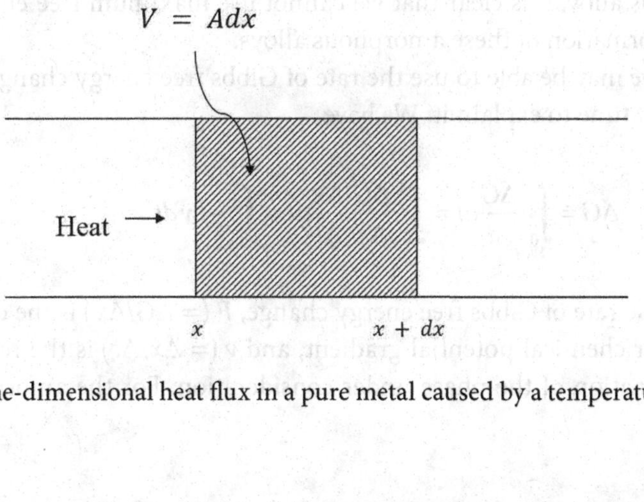

$$V = Adx$$

FIGURE 5.5 One-dimensional heat flux in a pure metal caused by a temperature gradient.

In the above, temperature is treated as a variable. Again, according to Eq. (5.8), the ratio of J_Q/T in the first team in the right-hand side of the above equation is an entropy flux, and $\frac{\partial}{\partial x}\left(\frac{J_Q}{T}\right)$ is a divergence of this flux, meaning the rate of entropy change per cell volume caused by the difference of incoming and outgoing entropy fluxes. The second term on the right-hand side of the above equation is the rate of entropy production per unit volume per unit time within the cell. This second term is a product of heat flux and the gradient of inverse temperature, which can be interpreted as the thermodynamic force (or the conjugate force) driving the heat flux. Because

$$\frac{\partial}{\partial x}\left(\frac{1}{T}\right) = -\frac{1}{T^2}\left(\frac{\partial T}{\partial x}\right)$$

Thus,

$$\frac{TdS}{Vdt} = -T\frac{\partial}{\partial x}\left(\frac{J_Q}{T}\right) - J_Q \frac{1}{T}\left(\frac{\partial T}{dx}\right) \tag{5.15}$$

In a steady-state process, the two terms on the right-hand side of the equation compensate each other. The first term is due to entropy divergence, and the second term is due to entropy production. We note that the dimension of the above equation is correct. It is energy/cm$^3 \cdot$ sec and we recall that the unit of heat flux J_Q is energy/cm$^2 \cdot$ sec [5].

5.4 RATE OF GIBBS FREE ENERGY CHANGE

After the discussion on entropy production given in the above, for comparison, we ask how about the rate of Gibbs free energy change for reactions in a closed system. For example, take the bilayer thin film reactions between Rd and Si or between Ni and Zr, where an amorphous alloy of Rd-Si or Ni-Zr was formed, respectively, upon slow heating rather than rapid quenching of the bilayer samples. Because the crystalline intermetallic compound phases of Rd-Si or Ni-Zr must have a lower free energy than the amorphous alloy, it is clear that we cannot use maximum free energy change to describe the formation of these amorphous alloys.

However, we may be able to use the rate of Gibbs free energy change (or gain) in a short period of time to explain it. We have

$$\Delta G = \int_0^\tau \frac{\Delta G}{\Delta t}dt = \int_0^\tau \frac{\Delta G}{\Delta x}\frac{\Delta x}{\Delta t}dt = \int_0^\tau -Fv\,dt \tag{5.16}$$

where G/t is the rate of Gibbs free energy change, F (= $\Delta G/\Delta x$) is the driving force of the reaction or chemical potential gradient, and v (= $\Delta x/\Delta t$) is the reaction velocity or rate of formation of the phase under consideration. For the amorphous alloy, its driving force of formation will be lower than that of the competing crystalline phases. But if its rate of formation (in a short period of time, τ) is greater than that of the crystalline phases, it is possible that the product of Fv for the amorphous alloy is greater than that of the crystalline phase, so the amorphous phase can form.

We can express Gibbs free energy change in a rate process as,

$$\frac{dG}{dt} = -S\frac{dT}{dt} + V\frac{dp}{dt} + \sum_i^j \mu_i \frac{dn_i}{dt} \qquad (5.17)$$

Indeed, experimentally, we can use rapid quenching (large change in dT/dt), rapid mechanical milling (large change in dp/dt), and ion implantation (large change in dn_i/dt) to produce amorphous or meta-stable materials. On the other hand, when time becomes infinitive, the equilibrium crystalline phase will win.

5.4.1 Crystallization of Amorphous Si

Furthermore, we consider the crystallization of an amorphous phase, such as amorphous Si. For a pure element, we have $C\Omega = 1$, where C is concentration and Ω is atomic volume. Because $J = C\langle v \rangle$, then in Eq. (5.16), we have

$$\Delta G = \Omega \int (-FJ)dt$$

The last equation show that we can have rate of Gibbs free energy change in terms of the product of driving force and atomic flux,

$$\Delta G/\Delta t = -\Omega(FJ)$$

which shows the link between the rate of Gibbs free energy change and the rate of entropy production.

However, there is a big difference between them because the crystallization or phase change of amorphous Si involves latent heat besides Joule heating. No doubt it is Joule heating that causes temperature increases and leads to crystallization. Yet, because of latent heat, we cannot just consider entropy production. Clearly, the electrical conduction of amorphous Si under a low current density without crystallization can be treated the same as that in the electrical conduction of a Cu wire, where only Joule heating occurs.

When we consider an ideal system, where enthalpy is zero, or $\Delta H = 0$, we have $\Delta G = -T\Delta S$. Thus, the rate of Gibbs free energy change is equal to the rate of entropy production. When we consider electrical conduction in a Cu wire, the Cu wire can be taken as an ideal system!

In Chapter 10, when we discuss the mean-time-to-failure equations for electromigration, thermomigration, and stress migration, we shall derive these equations based on entropy production.

5.5 CROSS-EFFECTS IN IRREVERSIBLE PROCESSES

Both electromigration and thermomigration are cross-effects in irreversible processes because they are atomic fluxes driven, respectively, by electron flow under an electrical potential gradient or by heat flow under a temperature gradient. On the other hand, stress migration is creep, and it is a flux of atoms driven by a stress potential gradient. Thus, creep does not occur under a uniform pressure or hydrostatic pressure; because it occurs under a stress potential gradient, it is an irreversible process too. However, the stress potential gradient is part of the chemical potential gradient that drives atomic diffusion, so creep is a primary flow, not a cross-effect.

The fundamentals of electromigration, thermomigration, and stress migration will be presented in later chapters. Next, we shall consider the cross-effects among atomic flow, heat flow, and charge flow.

5.6 CROSS-EFFECT BETWEEN ATOMIC DIFFUSION AND ELECTRICAL CONDUCTION

In Chapter 3, Figure 3.1 depicts a schematic diagram of a short Al strip patterned on a baseline of TiN. Under an applied high current density of electrons going from the cathode to the anode, or from left to right in Figure 3.1, electromigration occurs in the Al stripe, leading to depletion or void formation in the cathode and pile-up or hillock formation in the anode. The depletion rate in the cathode can be measured, and the drift velocity of electromigration can be deduced for the calculation of effective charge number. Furthermore, it was found that the longer the strip, the longer the depletion on the cathode side during electromigration, as shown in Figure 3.3. But below a "critical length," there was no observable depletion. The dependence of depletion on stripe length was explained by the effect of back stress, which has been discussed in Chapter 3, so it will not be repeated here.

5.7 CROSS-EFFECT BETWEEN ATOMIC DIFFUSION AND HEAT CONDUCTION

When a temperature gradient is applied to a homogeneous alloy, the alloy becomes inhomogeneous with phase separation. This is known as the Soret effect. One of the components in the alloy will diffuse against its concentration gradient, which leads to phase separation. Finally, it establishes a concentration gradient and reaches a steady state. Here, we shall consider the interaction between atomic flow and heat flow, and the corresponding conjugate forces are X_M and X_Q as shown in Eqs. (5.18a) and (5.18b), respectively. J_M is atomic flow and J_Q is heat flow.

$$J_M = L_{MM} X_M + L_{MQ} X_Q = C \frac{D}{kT}\left[-T\frac{d}{dx}\left(\frac{\mu}{T}\right)\right] + C\frac{D}{kT}\frac{Q^*}{T}\left(-\frac{dT}{dx}\right) \qquad (5.18a)$$

$$J_Q = L_{QM} X_M + L_{QQ} X_Q = L_{QM}\left[-T\frac{d}{dx}\left(\frac{\mu}{T}\right)\right] + \kappa\frac{dT}{dx} \qquad (5.18b)$$

where Q^* is the heat of transport, which is positive when the flux of atoms is diffusing from the hot end to the cold end because dT/dx is negative, and it is negative when atoms are driven from the cold end to the hot end because dT/dx is positive. The last term in the equation of J_Q is Fourier's law and κ is the thermal conductivity.

In the equation of J_M, if we let $J_M = 0$, we obtain

$$-T\left(\frac{1}{T}\frac{d\mu}{dx} - \frac{\mu}{T^2}\frac{dT}{dx}\right) = \frac{Q^*}{T}\frac{dT}{dx}$$

Rearrange the terms. We have

$$\frac{d\mu}{dT} = \frac{\mu - Q^*}{T}$$

Experimentally, to detect thermomigration by using the temperature gradient induced by Joule heating, a set of 24 bumps on the peripheral of a Si chip was tested. Figure 5.6a depicts a row of 24 solder bumps from right to left at the peripheral of a chip, and each bump has the original microstructure as shown in Figure 5.6b before electromigration stressing is discussed below. We recall that the darker region in the bottom area of each bump is the eutectic SnPb, and the brighter region in the top part is 97Pb3Sn.

Electromigration was conducted through only four pairs of bumps in the row of 24 bumps on the peripheral of the chip. They were the pairs of No. 6/7, 10/11, 14/15, and 18/19, as numbered in Figure 5.6a. The arrows indicated the electron path. The electron current went from one of the contact pads to the bottom of one of the bumps, up the bump to the Al thin film interconnect on the Si stripe, then went to the top of the next bump, down the bump, and to the other contact pad on the substrate. It is worth noting that we can pass current through just one pair of bumps or several pairs in a

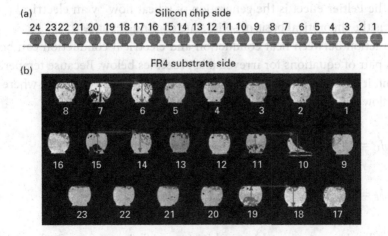

FIGURE 5.6 SEM images exhibiting the powerd and unpowered eutectic SnPb solder joints. (a) A row of 24 solder bumps from right to left at the peripheral of a chip. (b) Original microstructure of solder bumps before electromigration stressing.

row to conduct electromigration. The Joule heating from the Al thin film line on the stripe is the heat source. Due to the excellent thermal conduction of Si, the neighbour-ing un-powered solder joints should have experienced a thermal gradient similar to the pairs stressed by current.

A cross-sectional examination was performed after the bump-pair 10 and 11 failed after 5 hours of current stressing at 1.6×10^4 A/cm^2 at 150°C. To study thermomigration, those un-powered neighbouring bumps were examined. The effect of thermomigration is clearly visible across the entire row of the un-powered solder bumps, as shown in Figure 5.6b. Because in all of them, Sn has migrated to the Si side, the hot end, and Pb has migrated to the substrate side, the cold end. The redistribution of Sn and Pb, or the redistribution of the eutectic phase and the high-Pb phase, was caused by the tempera-ture gradient across the solder joints because no current was applied to them.

For the un-powered bumps, which were the nearest neighbours, of the powered bumps, the Sn redistribution is also tilted towards the powered bumps. For example, the powered bump 10 is to the left of the un-powered bump 9, and the Sn-rich region in bump 9 is tilted to the left, and a void is observed. Then, the powered bump 15 is to the right of the un-powered bump 16, and the Sn-rich region in bump 16 is tilted to the right. In those bumps further away from the powered bumps, for example, from bump 1 to bump 4 and bump 21 to 23, Sn accumulated rather uniformly on the Si side, which is the hot end.

5.8 CROSS-EFFECT BETWEEN ELECTRICAL CONDUCTION AND HEAT CONDUCTION

The cross-effect between heat flow and electrical flow is the thermal-electrical effect, especially the Seebeck effect and the Peltier effect, which are well known. The Seebeck effect is the generation of an electrical flow or electrical potential by a temperature gradient. It is the basic application of a thermal couple to measure the temperature of materials. The Peltier effect is the generation of a heat flow by an electrical potential gradient. It is the theoretical basis of solid-state cooling devices.

The interaction between heat conduction and electrical conduction can be repre-sented by a pair of equations for irreversible processes below. Because temperature is not constant, it becomes a variable in the forces in these two equations, where J_Q and J_E are heat flow and charge flow, respectively.

$$J_Q = L_{QQ} X_Q + L_{QE} X_E = L_{QQ} T \frac{d}{dx} \left(\frac{1}{T} \right) - L_{QE} T \frac{d}{dx} \left(\frac{\phi}{T} \right)$$

$$J_E = L_{EQ} X_Q + L_{EE} X_E = L_{EQ} T \frac{d}{dx} \left(\frac{1}{T} \right) - L_{EE} T \frac{d}{dx} \left(\frac{\phi}{T} \right)$$

(5.19)

where $L_{QQ} = T\kappa$, $L_{EE} = ne\mu_e$, and L_{QE} and L_{EQ} are coefficients of cross-effects. L_{QE} is for heat flow induced by an electrical field, and L_{EQ} is for the electrical flow induced by a temperature gradient.

On the basis of Onsager's reciprocity relationship, $L_{QE} = L_{EQ}$. We shall analyse the above two equations for the understanding of thermal-electric effects. We can rewrite the pair of equation in Eq. (5.19) in terms of dT/dx and $d\phi/dx$ as follows:

$$J_Q = (-L_{QQ} + \phi L_{QE})\frac{1}{T}\left(\frac{dT}{dx}\right) - L_{QE}\frac{d\phi}{dx}$$

$$J_E = (-L_{EQ} + \phi L_{EE})\frac{1}{T}\left(\frac{dT}{dx}\right) - L_{EE}\frac{d\phi}{dx} \quad (5.20)$$

5.8.1 Seebeck Effect

Figure 5.7a depicts a single metal wire of a given length, with its two ends kept at two temperatures, T_1 and T_2, where $T_1 > T_2$. Thus, there is a temperature gradient and we expect it to drive an electrical flow of charge carriers. Because the wire is open-ended, no electric flow will occur. Thus, in the second equation in Eq. (5.20), we have $J_E = 0$ and

$$0 = (-L_{EQ} + \phi L_{EE})\frac{1}{T}\left(\frac{dT}{dx}\right) - L_{EE}\frac{d\phi}{dx}$$

Then, we have

$$\frac{\Delta\phi}{\Delta T} = \frac{(-L_{EQ} + \phi L_{EE})}{TL_{EE}} \quad (5.21)$$

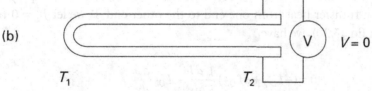

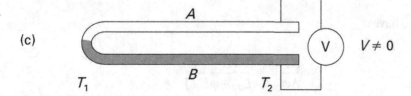

FIGURE 5.7 Schematic diagram of Seebeck effect (a) A single metal wire of a given length with its two ends kept at two temperatures T1 and T2. (b) Double the length of the wire and bend it at the middle into two branches and place the bent end at T1 and the other two ends at T2. (c) The bent wire with two different metal branches[6].

This is called the Thomson effect. It means that there will be an electrical potential difference between the ends due to the temperature gradient. In the above, we obtain $\Delta\phi/\Delta T$ by assuming $J_E = 0$.

Now we double the length of the wire, bend it in the middle into two branches, and place the bent end at T_1 and the other two ends at T_2, as shown in Figure 5.7b. While there will be an electric potential difference between the end at T_1 and the two ends at T_2, there will be no net potential difference between the two ends kept at T_2 because the potential changes in both branches are the same.

However, if we replace one branch of the bent wire with a different metal, or if we take two kinds of metallic wires, A and B, and join them at the end kept at T_1, as shown in Figure 5.7c, we have a thermal couple. If we place the joined end at a high temperature and the un-joined ends of the couple at a reference temperature of 0°C or room temperature, we have a thermal couple, and we can measure the potential difference $\Delta\phi$ between the two ends kept at the reference temperature. This is because the potential changes in the two wires are not the same, so a potential difference occurs. If we have calibrated the couple at different temperatures, we obtain the Seebeck coefficient,

$$\frac{\Delta\phi}{\Delta T} = \varepsilon_{AB} = \varepsilon_A - \varepsilon_B$$

where ε_{AB} is the combination of thermal-electric properties of A wire and B wire in the couple. For convenience, the thermal-electric properties of many pairs of individual material of ε_A and ε_B have been measured. So, we can choose a pair to serve as thermal couple to measure temperature in various temperature ranges.

5.8.2 Peltier Effect

The Peltier effect is opposite to the Seebeck effect. If we keep the two ends of the sample as shown in Figure 5.7c at two constant temperatures, and if we apply an electric field, we can transfer heat from one end to the other end. If we let $J_Q = 0$ in the first equation in Eq. (5.20), we have

$$0 = (-L_{QQ} + \phi L_{QE})\frac{1}{T}\frac{dT}{dx} - L_{QE}\frac{d\phi}{dx}$$

Thus, we have

$$\frac{\Delta T}{\Delta\varphi} = \frac{TL_{QE}}{(-L_{QQ} + \varphi L_{QE})} \tag{5.22}$$

In the above equation, ΔT is proportional to $\Delta\phi$ and L_{QE}. The larger the ΔT, the better the cooling effect. However, due to the existence of ΔT, heat will be transferred from the hot end to the cold end, which will reduce the cooling effect. In order to decrease heat transfer, we will need to reduce heat conductivity in the sample. Yet, for most conductors, heat conductivity is linearly proportional to electric conductivity. In applications of the Peltier effect, the challenge is to find a conductor that has good electrical conduction, but poor heat conduction.

PROBLEMS

5.1. In electromigration, what is the meaning of the effective charge number Z^*? For aluminium, how large is Z^*?

5.2. Why does electromigration in Al interconnect occur by grain boundary diffusion? But in the Cu interconnect, why does electromigration occur by surface diffusion?

5.3. Calculate the electron wind force Z^*eE at a current density of 10^5 A/cm^2 for Au and the stress potential of $\sigma\Omega$ at the elastic limit for Au, and calculate the critical length for Au.

5.4. Two Al short stripes with lengths of 20 μm and 30 μm undergo electromigration. Calculate the stress at their anode ends when they carry a current density if 10^5 A/cm^2?

5.5. What is the difference between reversible and irreversible processes?

REFERENCES

[1] I. Prigogine, *Introduction to Thermodynamics of Irreversible Processes* (3rd edn.), Wiley-Interscience, New York (1967).
[2] D. V. Ragone, Nonequilibrium thermodynamics, in *Thermodynamics of Materials*, Wiley, New York (1995) (Chapter 8).
[3] P. Shewman, *Diffusion in Solids* (2nd edn.), TMS, Warrendale, PA (1989).
[4] R. W. Balluffi, S. M. Allen, and W. C. Carter, Irreversible thermodynamics: coupled forces and fluxes, in *Kinetics of Materials*, Wiley-Interscience, New York (2005) (Chapter 2).
[5] J. C. M. Li, "Caratheodory's principle and the thermodynamic potential in irreversible thermodynamics", *J. Phys. Chem.*, 66, 1414–1420 (1962).
[6] K. N. Tu, "Electromigration in stressed thin films," *Phys. Rev.*, B45, 1409–1413 (1992).

CHAPTER 6

Effect of Stress on Electromigration

6.1 INTRODUCTION

Conceptually, there is a fundamental difference between stress migration and electromigration or thermomigration. The latter two are due to cross-effects based on irreversible processes, as discussed in Chapter 5. The difference is because the atomic flow in electromigration or thermomigration is accompanied by electron flow or heat flow, respectively. However, there is no "stress flow" to accompany the atomic flow in stress migration, especially if elastic stress is assumed. Therefore, stress migration is a primary flow of atoms, driven by stress potential energy gradient, which is also a chemical potential gradient. Stress migration is often called steady-state diffusional creep. We notice that there are many types of mechanical creeps, yet we only deal with diffusional creep here, based on the Nabarro-Herring creep model.

Stress potential is defined as $\sigma\Omega$, where σ is stress and Ω is atomic volume. Here, we recall the ideal gas law of $pV = kT$, where the units of pV are the same as that in $\sigma\Omega$, and kT is energy. So, the driving force of stress migration is given as,

$$F = -\frac{d\sigma\Omega}{dx} \tag{6.1}$$

In turn, the atomic flux in stress migration is given below:

$$J = CMF = \frac{1}{\Omega}\frac{D}{kT}\left(-\frac{d\sigma\Omega}{dx}\right) = \frac{D}{kT}\left(-\frac{d\sigma}{dx}\right)$$

where in a pure metal we have taken $C\Omega = 1$.

DOI: 10.1201/9781003384281-6

An example of steady-state diffusional creep is the sagging of Pb pipes by their own weight under gravity in many very old buildings. Clearly, room temperature is a relatively high temperature for Pb, whose melting point is 327°C, so atomic diffusion at room temperature is sufficient for creep to occur in Pb over years. Furthermore, a modern application of creep is the use of a pure and well-annealed O-ring of Cu as a pressure seal in ultra-high vacuum systems, where the compressive stress under tightened screws on the O-ring will enable the diffusion or redistribution of a few Cu atoms to fill up any tiny gaps in the seal. In electronic packaging technology, an example of stress migration is the growth of Sn whiskers under a compressive stress gradient, to be discussed later.

The interaction or the coupling between stress and electromigration has been presented in Section 3.3 of Chapter 3; therefore, it will not be repeated here. Next, we shall begin with the definition of chemical potential in a stressed solid and the basic analysis of bi-axial stress in thin films.

6.2 CHEMICAL POTENTIAL IN A STRESSED SOLID

When a piece of solid, especially a piece of metal, is under a steady state of stress (or under constant loading), even if it is within the elastic limit, the solid will respond by a slow deformation or relaxation owing to diffusional creep. We note that because elastic strain energy is small as compared to chemical bonding energy, we shall deal with creep in pure metals only in the absence of chemical bonding effects below.

We have the change of Helmholtz free energy F as

$$dF = -SdT - pdV$$

At a constant temperature change, as in a room temperature creep, we have

$$p = -\frac{\partial F}{\partial V}$$

This equation can be interpreted to mean that stress (pressure) is an "energy density" (i.e., energy per unit volume), which is a very useful understanding in thermodynamics. Considering a given volume, the energy equals the energy density times the given volume. Thus, for an atomic volume, Ω, under a stress of p, we have

$$p\Omega = -\frac{\partial F}{\partial V}\Omega = -\frac{\partial F}{\partial\left(\dfrac{V}{\Omega}\right)} = -\frac{\partial F}{\partial N}$$

where N is the number of atoms in volume V. We note that the last term in the above equation is the definition of chemical potential, where the negative sign is used to indicate that a decrease in volume under pressure results in an increase in energy because pressure is a compressive stress which is negative.

Here, because chemical potential is defined by Helmholtz (or Gibbs) free energy per atom, the chemical potential change in a stressed solid is

$$\Delta\mu = \pm\sigma\Omega \tag{6.2}$$

where the positive or negative sign refers, respectively, to an external tensile or compressive stress.

In order to gain a quantitative feeling of the magnitude of $\sigma\Omega$, we consider a piece of Al stressed at the elastic limit (i.e., strain is 0.2%). We have Young's modulus for Al to be $Y = 6\times1,011\,dyne/cm^2$, so the stress is

$$\sigma = Y\varepsilon = 1.2\times109\,dyne/cm^2 = 1.2\times109\,erg/cm^3$$

To calculate the atomic volume of Al, we recall that Al has a face-centred cubic lattice with a lattice constant of 0.405 nm, and there are four atoms in a unit cell of $(0.405)^3\,nm^3$, or $0.602\times19^{23}\,atoms/cm^3$. So, the stress potential energy is

$$\sigma\Omega = \frac{1.2\times10^9\,erg}{0.602\times10^{23}\,atom} = 2\times10^{-14}\,erg/atom = 0.0125\,eV/atom$$

For comparison of this value to the value of strain energy per atom, we have

$$E_{elastic} = \int\sigma\,d\varepsilon = \frac{1}{2}Y\varepsilon^2 = \frac{1}{2}\times1.2\times10^9\,erg/cm^3\times\frac{2}{1,000} = 1.2\times10^6\,erg/cm^3 = 10^{-5}\,eV/atom$$

which is much smaller than the stress potential energy.

We note that the strain energy is the increase in energy per atom due to the strain. The stress potential energy is the energy change from adding or removing one atom from the stressed solid. The latter is much larger.

In thermally activated processes such as diffusion or creep, the chemical potential energy enters as an exponential factor. We consider below the case of Al stressed to its elastic limit at 400°C; we have $kT = 0.058\,eV$, and

$$\exp-\left(\frac{\sigma\Omega}{kT}\right) = \exp-\left(\frac{0.0125}{0.058}\right) = 1.23$$

Typically, creep can occur at a much lower stress (or $\sigma\Omega \ll kT$); instead, we can linearize the exponential term by taking,

$$\exp\left(-\frac{\sigma\Omega}{kT}\right)=1-\frac{\sigma\Omega}{kT}$$

6.3 INTRINSIC THERMAL STRESS IN THIN FILM DEPOSITION

In thin film deposition, it is known to have an intrinsic stress in a thin film deposited at room temperature. To understand the intrinsic thermal stress, we consider the deposition of an Al thin film on a thick quartz substrate at room temperature, as shown in Figure 6.1a. We tend to assume that there is no stress in the Al thin film. However, for Ni or refractory metals, which have a high melting point, there will be a small tensile residue stress in the deposited film. The basic reason for the tensile stress has been explained by the formation of grain boundaries during deposition and during growth of the film [1–4].

We note that those atoms across a grain boundary are separated more than the equilibrium separation. According to the definition of pair potential, a pair of atoms are in tension when they are apart from the equilibrium distance. But the residual tension is small, and it may change to compression upon heating.

If we raise the sample temperature to 400°C, as shown in Figure 6.1b, owing to the greater thermal expansion of Al than quartz, the sample will bend downward and the Al is under compression. Then, if we keep the sample at 400°C for a while, creep will occur in the Al to relax the bending, as shown in Figure 6.1c, and the quartz will become flat again. This is because 400°C is high enough for rapid diffusion to occur in the Al thin film. Furthermore, if we lower the temperature to 100°C, the Al film will shrink and be under tension, as shown in Figure 6.1d. Owing to the fact that the creep rate at 100°C is slow in Al, we can measure the bending curvature to determine the tensile stress in the Al film by using Stoney's equation, to be discussed below.

The thermal expansion co-efficients for Al and quartz in the above temperature range are about $\alpha = 25 \times 10^{-6}/C$ and $\alpha = 0.5 \times 10^{-6}/C$, respectively. Therefore, the thermal strain is about

$$\varepsilon = \Delta\alpha\Delta T = 25 \times 10^{-6} \times 300 = 0.75\%$$

Then, the thermal stress in Al is

$$\sigma = Y\varepsilon = 4.5 \times 10^{9}\ \text{dyne/cm}^{2}, \text{and also } \sigma\Omega = 0.045\,\text{eV}.$$

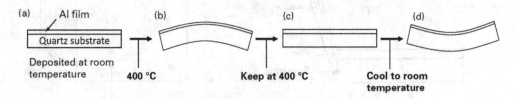

FIGURE 6.1 Process diagram of the experiment to explain the intrinsic thermal stress. (a) The deposition of an Al thin film on a thick quartz substrate at room temperature. (b) The sample bends downward (c) The sample creeps to relax the bending. (d) The Al film under tension.

6.4 STONEY'S EQUATION OF BI-AXIAL STRESS IN THIN FILMS

The stress is bi-axial in the Al thin film on quartz, as mentioned above [4]. As shown in Figure 6.2a, the stresses act along the two principal axes in the plane of the film, but most importantly, there is no stress in the direction normal to the free surface of the film. However, there can be strain in the normal direction.

To determine the bi-axial stress, we start with a three-dimensional isotropic cubic structure, as shown in Figure 6.2b. The linear dimensions in the x, y, and z axes are l, w, and t, respectively. We begin by considering the bulk cubic structure under hydrostatic compression, and then we shall reduce it step by step to the bi-axial compressive stress in a thin film.

We apply the compression in the x, y, and z directions in sequence.

Firstly, we apply the pressure p in the x-direction, and therefore

$$p = -Y \frac{\Delta l_1}{l}$$

Thus, the strain in the x-direction is

$$\frac{\Delta l_1}{l} = -\frac{p}{Y}$$

Secondly, we apply the compression in the y-direction, and we have

$$\frac{\Delta w}{w} = -\frac{p}{Y}$$

However, due to Poisson's effect, and by taking v as Poisson's ratio, we have the tensile strain in the x-direction owing to Poisson's effect given by

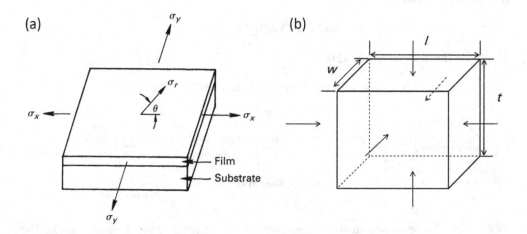

FIGURE 6.2 Biaxial stress in thin films. (a) The stresses act along the two principal axis in the plane of the film. (b) Three-dimensional isotropic cubic structure to express the biaxial stress.

$$\frac{\Delta l_2}{l} = +v\frac{p}{Y} = -v\frac{\Delta w}{w}$$

Thirdly, we apply the compression in the z-direction, and again, we have the tensile strain in the x-direction due to Poisson's effect given by

$$\frac{\Delta l_3}{l} = +v\frac{p}{Y}$$

The total strain in the x-direction is given by the sum of

$$\frac{\Delta l}{l} = \frac{\Delta l_1}{l} + \frac{\Delta l_2}{l} + \frac{\Delta l_3}{l} = -\frac{p}{Y}(1-2v)$$

Or $\quad \varepsilon_x = -\left(\dfrac{\sigma_x}{Y} - v\dfrac{\sigma_y}{Y} - v\dfrac{\sigma_z}{Y}\right) = -\dfrac{1}{Y}[\sigma_x - v(\sigma_y + \sigma_z)]$

Now we change the compressive stress to tensile stress, and we have the following equations:

$$\varepsilon_x = \frac{1}{Y}[\sigma_x - v(\sigma_y + \sigma_z)]$$

$$\varepsilon_y = \frac{1}{Y}[\sigma_y - v(\sigma_x + \sigma_z)]$$

$$\varepsilon_z = \frac{1}{Y}[\sigma_z - v(\sigma_x + \sigma_{zy})]$$

In the thin film bi-axial stress state, we assume there is tensile stress within the plane of the film (x and y-directions), but no stress in the z-direction or the normal direction ($\sigma_z = 0$). Therefore,

$$\varepsilon_x = \frac{1}{Y}(\sigma_x - v\sigma_y)$$

$$\varepsilon_y = \frac{1}{Y}(\sigma_y - v\sigma_x) \qquad\qquad (6.3)$$

$$\varepsilon_z = \frac{v}{Y}(\sigma_x - v\sigma_y)$$

From these equations, we have

$$\varepsilon_x + \varepsilon_y = \frac{1-v}{Y}\left(\sigma_x + \sigma_y\right)$$

$$\varepsilon_z = \frac{1}{1-v}\left(\varepsilon_x + \varepsilon_y\right)$$

Now, in two-dimensional isotropic systems, where $\varepsilon_x = \varepsilon_y$, we obtain

$$\varepsilon_z = \frac{2v}{1-v}\varepsilon_x$$

$$\sigma_x = \left(\frac{Y}{1-v}\right)\varepsilon_x \qquad (6.4)$$

Later on, we shall apply the relations in Eq. (6.4) in order to obtain Stoney's equation for the bi-axial stress of a thin film on a substrate [5].

To begin the analysis, we assume that the film thickness t_f is much smaller than that of the substrate thickness of t_s, so that the neutral plane, where there is no stress, can be taken to be in the middle of the substrate. In Figure 6.3a, we enlarge one end of the substrate in order to show the neutral plane, the stress distribution in the film and in the substrate, as well as the corresponding forces and moments. At equilibrium, the moment produced by the stress in the film must be equal to the moment produced by the stress in the substrate, as shown in Figure 6.3b.

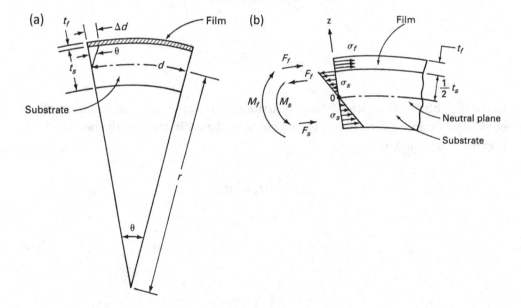

FIGURE 6.3 Schematic diagram of Stoney's equation. (a) The neutral plane, the stress distribution in the film and substrate. (b) The corresponding forces and moments.

Because we have assumed that the film thickness is thin, we can take the stress σ_f to be uniform across the entire film thickness. The moment M_f (force times the perpendicular distance) due to the force in the film with respect to the neutral plane is

$$M_f = \sigma_f W t_f \frac{t_s}{2} \tag{6.5}$$

where W is the width of film normal to t_f.

Next, we calculate the moment of the substrate, and we apply the following geometrical relation that

$$\frac{d}{r} = \frac{\Delta d}{t_s / 2}$$

$$\text{So,} \quad \frac{1}{r} = \frac{\Delta d}{d t_s / 2} = \frac{\varepsilon_{max}}{t_s / 2} \tag{6.6}$$

where r is the radius of curvature of the substrate measured from the neutral plane, d is an arbitrary length of the substrate measured at the neutral plane, and $\Delta d / d = \varepsilon_{max}$ is the strain at the outer surface of the substrate.

Within the substrate, the elastic strain is zero at the neutral plane; however, it increases linearly with the distance z, which is measured starting from the neutral plane (i.e., it obeys Hooke's law and increases linearly with stress), so that

$$\frac{\varepsilon_s(z)}{z} = \frac{\varepsilon_{max}}{t_s / 2} = \frac{1}{r}$$

where $\varepsilon_s(z)$ is the strain in a plane that is parallel to the neutral plane and is at a distance of z from the neutral plane. Therefore, by assuming a state of bi-axial stress in the substrate, we obtain from Eq. (6.4) that

$$\sigma_s(z) = \left(\frac{Y}{1-v}\right)_s \varepsilon_s(z) \quad \varepsilon_s(z) = \left(\frac{Y}{1-v}\right)_s \frac{z}{r} \tag{6.7}$$

Thus, the moment produced by the stress in the substrate is

$$M_S = w \int\limits_{-t/2}^{t_s/2} z\sigma(z)dz = w \int\limits_{-t_s/2}^{t_s/2} \left(\frac{Y}{1-v}\right)_s \frac{z^2}{r} dz = \left(\frac{Y}{1-v}\right)_s \frac{W t_s^3}{12r} \tag{6.8}$$

By equating M_S in Eq. (6.8) to M_f in Eq. (6.5), we obtain Stoney's equation as shown below:

$$\sigma_f = \left(\frac{Y}{1-v}\right)_s \frac{t_s^2}{6rt_f}$$ (6.9)

where the subscripts "f" and "s" refer to film and substrate, respectively.

Eq. (6.9) shows that by measuring the curvature and the thickness of the film and the substrate, and by knowing the Young's modulus and Poisson's ratio of the substrate, the bi-axial stress in the thin film can be determined by knowing the curvature r, which can be measured by methods of bending beam or laser interference. Knowing the stress distribution, we can now calculate the creep behaviour in thin films below.

6.5 DIFFUSIONAL CREEP

Diffusional creep is a time-dependent mechanical or deformation behaviour of solids. The deformation involves atomic diffusion when the applied load or stress is steady, within the elastic limit, but no plastic deformation occurs.

6.5.1 Nabarro-Herring Creep Model

In Figure 6.4a, we depict a square solid, where the length of each side is $2r$, and thickness is unity. Upon applying a steady-state compressive stress in the vertical direction, if the stress persists, the solid can be deformed slowly by transporting the atoms from the upper and lower shaded areas, which are the compressive regions, to go to both vertical sides where there is no normal stress. Transport can occur by atomic diffusion or, inversely, by vacancy diffusion. We shall present below the Nabarro-Herring creep model to analyse this problem. It assumes that surface and grain boundaries are effective sources and sinks of vacancies, which can mediate mass transport by atomic diffusion or vacancy diffusion in the lattice.

On the vertical side surfaces, because there is no normal stress, so

$$\mu_1 = 0$$

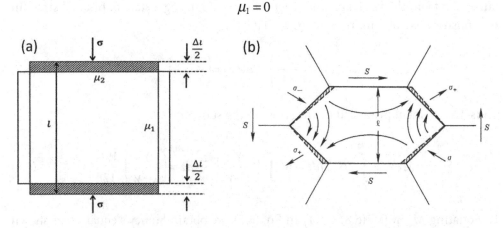

FIGURE 6.4   Schematic diagram of Nabarro-Herring creep model. (a) Deformation of a square solid under a steady state compressive stress. (b) A hexagonal grain in a polycrystalline solid under a shear stress.

In the compressive regions on the top and bottom surfaces, the chemical potential, which deviates from the equilibrium value, μ_0, due to the stress is given by

$$\mu_2 - \mu_0 = -\sigma\Omega$$

Thus, the chemical potential difference in going from the compressive region to the vertical side surfaces is

$$\Delta\mu = \mu_2 - \mu_1 = -\sigma\Omega$$

This potential difference will drive atoms to diffuse from the compressive regions to the tensile regions. The force acting on the diffusing atoms is given below:

$$F = -\frac{\Delta\mu}{\Delta x} = \frac{\sigma\Omega}{r}$$

where r is half the vertical size of the sample. The flux of the diffusing atoms can be given as

$$J = CMF = C\frac{D}{kT}\left(\frac{\sigma\Omega}{r}\right) = \frac{D\sigma}{kTr} \tag{6.10}$$

In the above, we have taken $C\Omega = 1$ for a pure metal.

The number of atoms transported by the flux of the diffusing atoms in a period "t" and through an area "A" is $N' = JAt$, and then the volume accumulated by the atomic flux is $\Omega N' = \Omega JAt$, where Ω is atomic volume. The strain is

$$\varepsilon = \frac{\Delta l}{l} = \frac{2\Omega JAt}{2r} = \frac{\Omega Jt}{r}$$

$$\frac{d\varepsilon}{dt} = \frac{\Omega J}{r} = \frac{\Omega D\sigma}{kTr^2}$$

This is the well-known Nabarro-Herring creep equation, where the strain rate has a dependence on $1/r^2$.

Next, we shall give one more example of the Nabarro-Herring creep model. In Figure 6.4b, we depict a hexagonal grain in a polycrystalline pure metal under a shear stress "S," which can be regarded as under a combination of tensile and compressive stresses, or vice versa. The effect of the shear stress, or the combined tensile and compressive stresses, is to deform the hexagonal grain from its original shape, delineated by the solid lines, to that delineated by the broken lines.

If the stress persists, the deformation occurs slowly by transporting the atoms in the shaded areas from the compressive region to the tensile region in order to release the stress. The transport occurs by atomic diffusion, as indicated by the curved arrows.

In the tensile region, the chemical potential, which deviates from the equilibrium value, μ_0, due to the stress is given by

$$\mu_1 - \mu_0 = \sigma\Omega$$

Similarly, in the compressive region,

$$\mu_2 - \mu_0 = -\sigma\Omega$$

Therefore, the chemical potential difference in going from the compressive region to the tensile region is given as

$$\Delta\mu = \mu_2 - \mu_1 = -2\sigma\Omega$$

This potential difference will generate a force to drive atoms to diffuse from the compressive regions to the tensile regions. The force is given as

$$F = -\frac{\Delta\mu}{\Delta x} = \frac{2\sigma\Omega}{l}$$

where l is the grain size. The flux of the diffusing atoms can be given below:

$$J = CMF = C\frac{D}{kT}\left(\frac{2\sigma\Omega}{l}\right) = \frac{2\sigma D}{kTl}$$

In the above, we have taken $C\Omega = 1$ for a pure metal.

The number of atoms transported by the flux of the diffusing atoms in a period "t" and through an area "A" is $N' = JAt$, then the volume accumulated by the atomic flux is $\Omega N' = \Omega JAt$. The strain is

$$\varepsilon = \frac{\Delta l}{l} = \frac{\dfrac{\Omega N'}{A}}{l} = \frac{\Omega Jt}{l}$$

So, the strain rate or the creep rate is

$$\frac{d\varepsilon}{dt} = \frac{\Omega J}{l} = \frac{2\sigma\Omega D}{kTl^2} \qquad (6.11)$$

Again, we obtain the well-known Nabarro-Herring creep equation. It shows an inverse dependence on l^2. To check the units in the above equation, we obtain

$$1/\sec = \left(F/cm^2\right)cm^3\left(D/kT\right)/cm^2 = \left(F/cm^2\right)cm^3\left(cm/\sec/F\right)/cm^2 = 1/\sec.$$

The units are correct in the equation.

6.5.2 Creep by Vacancy Diffusion

The above derivations are obtained by considering the flux of diffusing atoms in the lattice. However, if we assume lattice atomic diffusion occurs by vacancy mechanism, as in face-centred cubic metals of Al or Cu, we should be able to derive the same equation by considering the inverse flux of vacancies. To do so, we must deal with the change in concentration of vacancy in the tensile region as well as in the compressive region.

The formation energy of a vacancy in a face-centred cubic metal has been taken to be the energy needed to break all the 12 nearest-neighbour bonds of the atom, remove the atom, and place it on the metal surface. In the tensile region, the breaking of the bonds will be slightly easier because the tensile stress has stretched the bonds slightly, but in the compressive region, it will be slightly harder because the compressive stress has shortened the bonds slightly.

Thus, we can express the change in concentration of vacancy in a stressed solid as

$$C_v^{\pm} = C\exp[(-\Delta G_f \pm \sigma\Omega)/kT] \tag{6.12}$$

where C_v^+ and C_v^- correspond to the vacancy concentration in the tensile and compressive regions, respectively. Assuming $\sigma\Omega \ll kT$, we have

$$C_v^{\pm} = C_v\left(1 \pm \frac{\sigma\Omega}{kT}\right)$$

where $C_v = C\exp\left(-\Delta Gf/kT\right)$ is the concentration of vacancies in the equilibrium state or under no stress. Therefore, the vacancy concentration difference between the tensile region and the compressive region is

$$\Delta C_v = C_v^+ - C_v^- = C_v\frac{2\sigma\Omega}{kT} \tag{6.13}$$

which means the flux of vacancy going from the tensile region to the compressive region is

$$J_v = -D_v\frac{\Delta C_v}{\Delta x} = -\frac{2\sigma\Omega D_v C_v}{kTl} \tag{6.14}$$

where D_v is the diffusivity of vacancy, and $\Delta x = l$. Now, by taking $DC = D_v C_v$, we obtain

$$J = \frac{2\sigma\Omega DC}{kTl} = \frac{2\sigma D}{kTl} \tag{6.15}$$

which is the same as Eq. (6.10). So, whether we consider atomic diffusion or vacancy diffusion, the creep equation is the same.

The advantage of considering vacancy diffusion is that when creep occurs in a microstructure where there are interfaces, the vacancy flux may lead to void formation at the interfaces, provided that the nucleation of a void can occur. To calculate the void volume or rate of void growth, it will be easier to consider vacancy diffusion instead of atomic diffusion.

In Eq. (6.11), if we plot $\ln(Td\varepsilon/dt)$ versus $1/kT$, we can determine the activation energy of creep, which has been found to be the same as the activation energy of lattice diffusion in the solid. Indeed, this correlation has been verified in the literature.

6.6 SPONTANEOUS GROWTH OF SN WHISKERS AT ROOM TEMPERATURE

6.6.1 Morphology

Figure 6.5a is an SEM image of a Sn whisker formed between two Cu bumps. Both have a top layer of Sn, and the whisker is an electrical short between them. Figure 6.5b shows no Sn whisker when a Ni layer was added between the Cu bump and the Sn layer. The mechanism of spontaneous growth of Sn whiskers will be discussed below. Spontaneous growth means the growth occurs naturally at room temperature, like a bamboo shoot.

First, we consider the lead-frame (or leg-frame) in electronic packaging technology, where the Cu legs have a surface coating of Sn about 15 µm thick, which enables

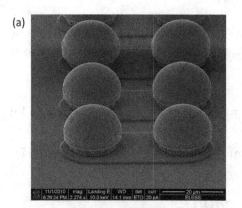
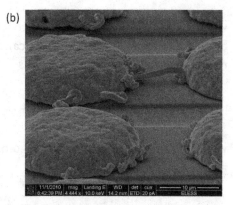

FIGURE 6.5 The morphology of Cu bump and whiskers. (a) SEM image without Sn whisker between the Cu bump and the Sn layer. (b) SEM image of a Sn whisker between two Cu bumps.

the spontaneous growth of Sn whiskers at room temperature, as shown in Figure 6.6a and b. The cross-sectional TEM image of Sn whiskers and the corresponding electron diffraction patterns are shown in Figure 6.7a and b, respectively.

Typically, whisker growth occurs on the surface of a bi-layer of Sn on Cu, where the thickness of Sn should be over 10 μm thick. Owing to the fact that Cu diffuses interstitially in Sn, as well as the fact that Cu and Sn can react at room temperature to form Cu_6Sn_5, they provide the unique driving force for spontaneous growth of Sn whiskers.

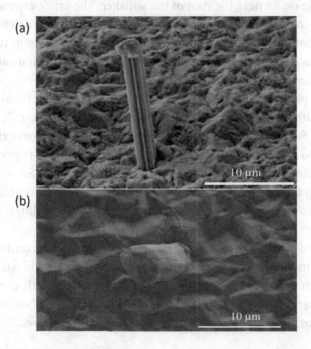

FIGURE 6.6 The morphology of Sn whisker. (a) SEM image of Sn whiskers on a lead-frame surface. (b) A short whisker of the lead-frame surface.

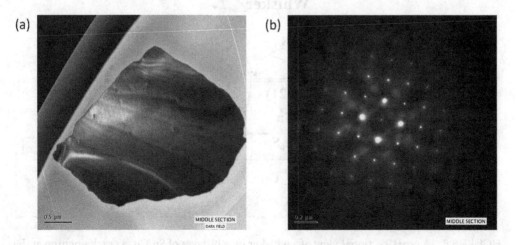

FIGURE 6.7 The morphology of Sn whisker. (a) The cross-sectional image of Sn whiskers. (b) The corresponding electron diffraction pattern, the growth orientation of the Sn whisker is (001) [6].

The growth of Cu_6Sn_5 along the grain boundaries in Sn generates a compressive stress in the surrounding region of Sn. The stress is compressive because Cu atoms are added to a fixed volume of Sn. To release the compressive stress, some of the Sn atoms must diffuse away, but they will need a stress gradient as the driving force.

It is worth noting that the growth of a whisker exerts a pulling force at the root of the whisker or at the interface between the whisker and its substrate. Due to the fact that the whisker is covered completely by Sn oxide, the pulling can induce a crack in the surface oxide on Sn near the root of the whisker. The crack exposes a free surface of Sn, which is stress-free. Therefore, there is a stress gradient to drive the Sn atoms to diffuse to the root of the whisker, so the whisker can grow at room temperature. We recall that Sn has a low melting point at 232°C, so atomic diffusion can occur in Sn at room temperature along its grain boundaries.

Figure 6.8 depicts the cross-sectional view of a whisker on a bi-layer of Sn on Cu. An arrow indicates the crack opening at the bottom of the whisker. The crack exposes a free surface of Sn, which has no normal stress, so a stress gradient exists between the crack surface and the compressive region in Sn. The compressive region is indicated by a square of broken lines in Figure 6.8. The stress gradient drives Sn atoms to diffuse to the surrounding area of the crack to grow the whisker. The stress gradient is actually rather small, measured by synchrotron radiation micro X-ray diffraction, which will be discussed later.

We can regard whisker growth as a room temperature recrystallization in which the Sn under compressive stress transforms to a stress-free Sn whisker.

In the literature, it has been reported that adding rare earth elements to solder enhances whisker growth. This is because rare earth elements are strong oxidation elements; their oxide helps crack the Sn oxide and whisker follows.

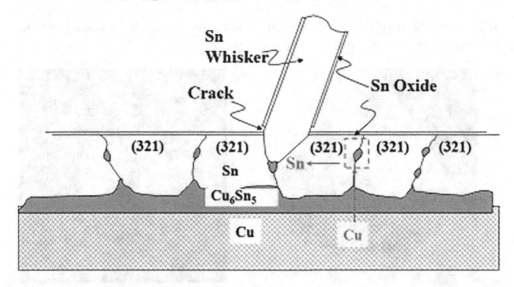

FIGURE 6.8 A cross-sectional view of a whisker on a bi-layer of Sn/Cu, a crack opening at the bottom of the whisker is indicated by an arrow.

Often, Sn whiskers are not straight, but they bend. This can be explained by the partial oxide formation at the whisker root. If a part of the oxide surrounding the root is not broken, the whisker growth stops on this part. Then the growth on the opposite side will lead to a tilting of the whisker. However, when the whisker has bended to 90°, it cannot bend more, and a straight growth will resume.

Even though we understand the basic mechanism of whisker growth, achieving no whisker growth in a device remains very challenging. This is because it is hard to guarantee that not even one whisker is allowed to grow in a device, such as satellite, which requires very high reliability. This is why the research on whisker growth is still active!

The condition of being completely free of Sn whisker growth is difficult to meet, except if we can get rid of Sn-based solder joints or cover the entire surface of the device with a strong coating. The Ni layer in Figure 6.5b is to block the diffusion of Cu into Sn to prevent the growth of Cu_6Sn_5 and, in turn, the growth of Sn whiskers. Furthermore, we recall that it is rare to find Sn whisker growth on high-Pb Sn-Pb solder. However, due to the poisoning effect of Pb on the environment, only server computers or military computers can use Pb-containing solder.

In the next section, we shall present a direct measurement of the stress gradient near the root of a whisker, which turns out to be quite small. Combining the fast kinetics of interstitial diffusion of Cu in Sn as well as the rather fast self-diffusion of Sn in Sn at room temperature, we can understand why Sn whisker growth is spontaneous and why it is a unique metallurgical phenomenon.

6.6.2 Driving Force of Sn Whisker Growth: Measurement of the Compressive Stress to Grow a Sn Whisker

In Figure 6.9, an SEM image of Sn whiskers on a lead-frame surface is shown. The circle depicts the root area of the Sn whisker to be studied by synchrotron radiation micro X-ray

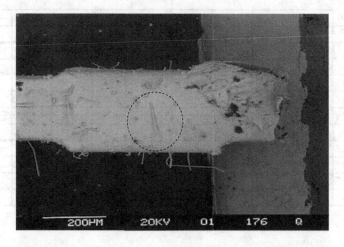

FIGURE 6.9   SEM image of a Sn whisker on a leg of lead-frame indicated by a circle.

diffraction. The image of the tip of the whisker is out of focus because the whisker sticks out. In Figure 6.10, the area surrounding the Sn whisker (in red) to be measured by synchrotron radiation micro X-ray diffraction is shown. In Figure 6.11, the measured distribution of compressive stress in the surrounding area around the root of the whisker is shown.

We can estimate in Figure 6.11 the stress gradient between the origin $(x = 0, y = 0)$ which is at the root of the whisker and the lower left corner point at

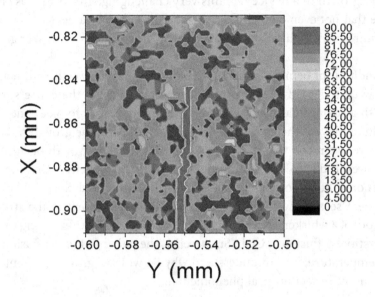

FIGURE 6.10   Synchrotron radiation micro-diffraction diagram measuring the stress distribution around the root of the whisker, the x and y coordinates enabled us to locate the position precisely.

|1.5 µm| | | | | | | | | | (Unit:MPa)|
	-0.5400	-0.5415	-0.5430	-0.5445	-0.5460	-0.5475	-0.5490	-0.5505	-0.5520	-0.5535	-0.5550
-0.8340	-2.82	-3.21	-2.26	0.93	0.93	-0.23	-8.17	2.22	1.49	1.6	-0.03
-0.8355	-2.26	-2.64	-2.64	-1.04	1.37	1.37	-1.31	0.87	0.87	0.87	-0.7
-0.8370	-2.53	-3.21	-3.21	-2.64	-1.04	3.61	0.75	0.87	0.7	0.7	-0.19
-0.8385	-7.37	-9.62	-6.57	-2.64	3.61	4.52	3.61	0.29	-1.31	0	-4.79
-0.8400	-7.37	-8.22	-6.57	-1.18	0.75	4.23	0.75	-2.25	-2.27	-2.91	-6.91
-0.8415	-4.17	-4.84	-4.17	-1.81	-0.67	2.00	-1.96	-1.96	-3.74	-5.08	-5.08
-0.8430	-4.17	-4.17	-3.63	-1.81	-1.81	-2.29	-2.29	-1.96	-1.96	-3.27	-3.27
-0.8445	-4.14	-4.17	-3.86	-3.63	2.79	-4.64	-4.78	-0.84	-1.4	-1.49	-3.27
-0.8460	-3.14	-3.63	-3.86	-3.63	-3.13	-4.78	-4.78	0.04	0.04	-1.41	-2.33
-0.8475	-4.14	-4.49	-4.49	-4.64	-3.86	-6.04	-1.72	3.55	3.55	-0.41	-2.33
-0.8490	-3.33	-5.67	-6.29	-6.29	-2.66	-2.08	-1.72	-1.79	0	-1.79	-3.73

Whisker

FIGURE 6.11   The measured distribution of compressive stress in the surrounding area of the root of the whisker [7].

$(x = -0.5400, y = 0.8475)$. We found that $\Delta x \approx 10\mu m$ and $\Delta\sigma \approx 4\,MPa$. When we take the atomic volume of Sn atom to be $27\times10^{-24}\,cm^3$, the driving force is obtained as shown below:

$$F = -\frac{\Delta\sigma\Omega}{\Delta x} = \frac{4\times10^6(N/m^2)\times27\times10^{-24}(cm^3)}{10^{-3}(cm)} = \frac{4\times10^7(dyne/cm^2)\times27\times10^{-24}(cm^3)}{10^{-3}\,cm}$$

$$= \frac{108\times10^{-17}\,erg}{10^{-3}\,cm} \approx 10^{-12}\,erg/cm$$

The work done by this force over a distance of atomic diameter of 0.3nm is 3×10^{-27} joule, which is of the same order of magnitude of those forces calculated for electromigration and thermomigration in Sn, which will be shown in the last section of this chapter. The small driving force needed to grow a whisker is one of the reasons why Sn whisker is so common.

Whisker growth has also been found to take place at the anode of a Sn-based solder joint in electromigration. Current crowding occurs at the anode of the solder joint, where a whisker was found to have been squeezed out, to be discussed in a later section.

6.6.3 Kinetics of Sn Whisker Growth

In Figure 6.12, the model of the kinetics of Sn whisker growth is shown. We assume an array of whiskers, each with a diameter of "$2a$," and the stress field that supplies atomic flux for the growth has a radius of "b." To evaluate the driving force in the stress field under cylindrical coordination, we assume

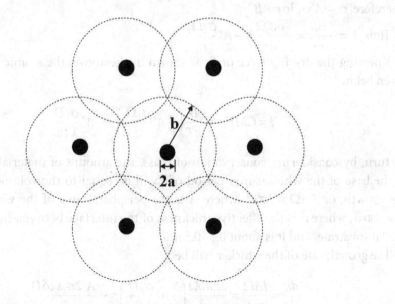

FIGURE 6.12 The model of the kinetics of Sn whisker growth[8]

$$F = -\frac{\partial \mu}{\partial r}$$

For simplicity, we just consider the early stage of growth, assuming a steady state. Therefore, we solve in a cylindrical coordinate, two-dimensional, and stead-state continuity equation of the potential $\sigma \Omega$.

$$\nabla^2 \sigma = \frac{\partial^2 \sigma}{\partial r^2} + \frac{1}{r}\frac{\partial \sigma}{\partial r} = 0$$

The reason why we can use the continuity equation to solve "σ" is because it is a density-like function, which has been discussed in the beginning of this chapter.

The general solution has the form of $\sigma = A'' \ln r + B''$. By using the boundary condition at t (time) $= 0$, which is taken to be the very beginning of whisker growth, we have

$$\sigma = \sigma_0, \text{where } r = b.$$

$$\sigma = \sigma, \text{where } r = a.$$

and we obtain

$$A'' = \frac{\sigma_0}{\ln(b/a)} = \sigma_0 A' \quad B'' = -\frac{\sigma_0 \ln a}{\ln(b/a)}$$

Therefore, $\sigma = A' \sigma_0 \ln r + B''$

Thus, $F = -\frac{\partial \mu}{\partial r} = -\frac{\partial \sigma \Omega}{\partial r} = -A' \frac{\sigma_0 \Omega}{r}$

Knowing the driving force of F, as shown in the above, the atomic flux at $r = a$ is given below:

$$J = CMF = C\frac{D}{kT}\left(-A'\frac{\sigma_0 \Omega}{a}\right) = -A'\frac{\sigma_0 D}{kTa}$$

In turn, by considering conservation of mass, the amount of material accumulated at the base of the whisker in a period of t will be equal to the volume of the whisker growth, or $JАt\Omega = h\pi a^2$, where A is the peripheral area of the whisker base, so $A = 2\pi r \delta$, where δ is the effective thickness of the interface between the whisker and the Sn substrate, and it is about 0.3–0.5 nm.

The growth rate of the whisker will be

$$\frac{dh}{dt} = \frac{JA\Omega}{\pi a^2} = \frac{2\pi a \delta \Omega}{\pi a^2}\left(-A'\frac{\sigma_0 D}{kTa}\right) = -\frac{A' 2\sigma_0 \Omega \delta D}{kTa^2}$$

where D is the diffusivity along the interface of width δ. The final height of the whisker, "h_f," which can be estimated again by using the law of conservation of mass, and

$$\pi a^2 h_f = \pi b^2 h' \varepsilon$$

$$\text{or} \quad h_f = \frac{b_2 h'}{a^2} \sigma_0 \frac{(2-3v)}{Y}$$

where h' is the thickness of the Sn film and $\varepsilon = \sigma_0(2-3v)/Y$ is the strain of the film, and v is Poisson's ratio and Y is Young's modulus. The estimate rate and the height of whisker are in reasonable agreement with experimental data.

6.7 SN WHISKER GROWTH IN SOLDER JOINTS INDUCED BY ELECTROMIGRATION

In Figure 6.13a–d, a set of four SEM images shows the initiation as well as the subsequent growth of a Sn whisker at the anode end of the Pb-free solder joint under electromigration. The flux of electrons flew in from the lower left corner and exited from the upper right corner, as indicated by the long arrow. It shows that electromigration is driving Sn atoms to the anode end at the upper right corner. Because the surface of Sn is covered by Sn oxide, which is protective, compressive stress is built up at the anode.

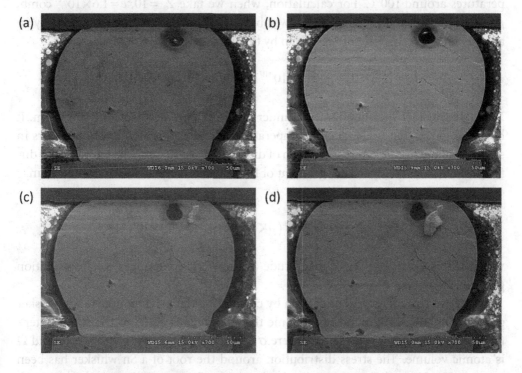

FIGURE 6.13 A set of four SEM images of the initiation and growth of a Sn whisker at the anode side of a Pb-free solder joint in electromigration from 0-240 h (a-d) [9].

The compressive stress cracked the surface oxide, see Figure 6.13b, so a small area of free surface is formed, which is stress-free. Hence, a stress potential gradient is produced between the stress-free broken surface and the surrounding anode, which is under compression. The stress potential gradient enables stress migration of Sn to occur and leads to Sn whisker growth, see Figure 6.13c and d.

No doubt, the whisker must have Sn oxide on the surface, yet the growth of the whisker can stretch and break the surface oxide near its root. While electromigration maintains the compressive stress, the broken oxide maintains the stress-free surface. In combination, whisker growth occurs. We emphasize here that broken oxide is a key condition in Sn whisker growth.

6.8 COMPARISON OF DRIVING FORCES AMONG ELECTROMIGRATION, THERMOMIGRATION, AND STRESS MIGRATION IN SN

A comparison is made below among the driving forces of electromigration, thermo-migration, and stress migration in Sn, especially the growth of Sn whiskers under a compressive stress. We expect that they should be of the same order of magnitude.

First, we consider electromigration, where the driving force is $F = Z^* eE = Z^* e\rho j$. Furthermore, it is known that electromigration occurs in Pb-free and Sn-based solder joints when the applied current density is above 10^4 A/cm^2 or 10^8 A/m^2 at temperatures around 100°C. For calculation, when we take $Z^* = 10$, $e = 1.6 \times 10^{-19}$ comb, $\rho = 10 \times 10^{-8}$ Ωm, and $j = 1 \times 10^8$ A/m^2, we obtain $F = 1.6 \times 10^{-17}$ N. The work done by an atomic jump of a distance of "a" driven by this force will be

$$\Delta w = Fa = (1.6 \times 10^{-17} \text{ N})(3 \times 10^{-10} \text{ m}) = 4.8 \times 10^{-27} \text{ Nm} = 4.8 \times 10^{-27} \text{ J}.$$

Next, we consider thermomigration under a temperature gradient of 1,000°C/cm. It has been found that under such a temperature gradient, thermomigration occurs in solder joint near 100°C. Across an atom of diameter of 3×10^{-8} cm, the temperature difference due to the temperature gradient of 1,000°C/cm is about 3×10^{-5} K. The change of thermal energy across the atom is

$$3k\Delta T = 3 \times 1.38 \times 10^{-23} \text{ (J/K)} \times 3 \times 10^{-5} \text{ K} = 1.3 \times 10^{-27} \text{ J}$$

which is of the same order of magnitude as the work done under electromigration given in the above.

Lastly, we consider stress migration by calculating the driving force of Sn whisker growth. On stress migration, we assume that the force is equal to the stress potential gradient, which is $- d\sigma\Omega/dx$, where σ is normal stress (not shear stress) and Ω is atomic volume. The stress distribution around the root of a Sn whisker has been measured by X-ray diffraction using synchrotron radiation, as shown in Figure 6.11.

Here, we consider the stress gradient between the origin ($x = 0$, $y = 0$) and the lower left corner point at ($x = -0.5400$, $y = 0.8475$). We found that $\Delta x \approx 10\,\mu m$ and $\Delta \sigma \approx 4\,MPa$, and we take the atomic volume of the Sn atom to be $27 \times 10^{-24}\,cm^3$. The force is given below:

$$F = -\frac{\Delta \sigma \Omega}{\Delta x} = \frac{4 \times 10^6 (N/m^2) \times 27 \times 10^{-24}(cm^3)}{10^{-3}(cm)} = \frac{4 \times 10^7 (dyne/cm^2) \times 27 \times 10^{-24}(cm^3)}{10^{-3}\,cm}$$

$$= \frac{108 \times 10^{-17}\,erg}{10^{-3}\,cm} \approx 10^{-12}\,erg/cm$$

Therefore, the work done by this force over a distance of atomic diameter of $0.3\,nm$ is 3×10^{-27} joule, which is of the same order of magnitude as those calculated in the above for electromigration and thermomigration in Sn.

PROBLEMS

6.1. What is the difference between stress potential energy and strain energy?

6.2. In thin film deposition, typically, there is a residual tensile stress. Why?

6.3. Both electromigration and thermomigration are cross-effects in irreversible processes. How come stress migration is not a cross-effect?

6.4. Upon applying an elastic stress to a piece of metal, what is the primary flux or flow in the metal?

6.5. When we couple stress migration and electromigration, a critical length is obtained below which there is no electromigration. In comparison, when we couple stress migration and thermomigration, can we find a critical length or not? If not, why not?

REFERENCES

[1] N. F. Mott and H. Jones, *The Theory of the Properties of Metals and Alloys*, Dover, New York (1958).
[2] P. Chaudhari, "Grain growth and stress relief in thin films," *J. Vacuum Sci. Technol.*, 9, 520 (1972).
[3] W. D. Nix, "Mechanical properties of thin films," *Metall. Trans.*, A, 20, 2217–2245 (1989).
[4] F. Spaepen, "Interfaces and stress in thin films," *Acta Mater.*, 48, 31–42 (2000).
[5] M. Murakami and A. Segmiller, "Analytical techniques for thin films", ed. by K. N. Tu and B. Rosenberg, in *Treatise on Materials Science and Technology*, Academic Press, Boston, MA, Vol. 27 (1988).
[6] G. T. T. Sheng, C. F. Hu, W. J. Choi, K. N. Tu, Y. Y. Bong, and L. Nguyen, "Tin whiskers studied by focused ion beam imaging and transmission electron microscopy," *J. Appl. Phys.*, 92, 64–69 (2002).

[7] W. J. Choi, T. Y. Lee, K. N. Tu, N. Tamura, R. S. Celestre, A. A. MacDowell, Y. Y. Bong, and L. Nguyen, "Tin whisker studied by synchrotron radiation micro-diffraction," *Acta Mater.*, 51, 6253–6261 (2003).

[8] K. N. Tu, "Irreversible processes of spontaneous whisker growth in bimetallic Cu-Sn thin film reactions," *Phys. Rev.*, B49, 2030–2034 (1994).

[9] F.-Y. Ouyang, K. Chen, K. N. Tu, and Y.-S. Lai, "Effect of current crowding on whisker growth at the anode in flip chip solder joints," *Appl. Phys. Lett.*, 91, 231919 (2007).

CHAPTER 7

Effect of Current Crowding on Electromigration

7.1 INTRODUCTION

When a flow of electrons makes a turn in a three-dimensional microstructure (3D IC), current crowding occurs. Even in 2D IC circuits, via is used to connect wires in two levels; it is actually a 3D microstructure. Whether the flow of electrons goes from right to left or from down to up, all electrons take the shortest path during the turn in order to reduce resistance; therefore, it leads to current crowding.

In Figure 7.1, the phenomenon of current crowding is shown, where a Ni silicide line is formed under an applied high current density in the heavily doped Si channels. Figure 7.1a shows a straight silicide line in the Si channel because the cathode and the anode are aligned along a straight line. Figure 7.1b shows the silicide line

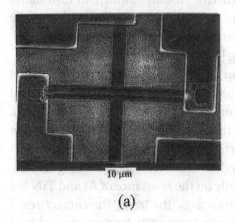
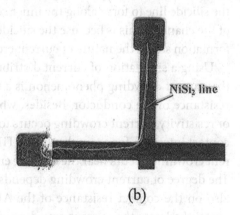

NiSi$_2$ line

10 μm

(a)   (b)

FIGURE 7.1 The phenomenon of current crowding. (a) A straight silicide line in the Si channel. (b) The silicide line makes a 90 degree turn.

DOI: 10.1201/9781003384281-7

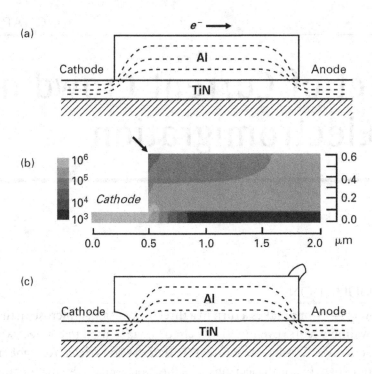

FIGURE 7.2 Schematic diagram of the Al/TiN interface in a short stripe of Al on TiN. (a) Current crowding near the cathode and anode end. (b) Current density distribution plot where the black arrow represents low current density at upper left corner. (c) The assumption of void growth at the lower left corner [6].

turns because the electrodes are at right angles to each other, making a 90° turn. However, what is unique in Figure 7.1b is the current crowding, which, in turn found the silicide line to form along the inner corner of the channel rather than in the middle of the channel. This is because the silicide formation was driven by electric current; the formation shows the nature of current crowding [1–5].

Using a simulation of current distribution in a conductor, we can easily show that the current crowding phenomenon is a function of geometry as well as the electrical resistance of the conductor. Besides, when a conductor changes its thickness, width, or resistivity, current crowding occurs too. Furthermore, when a current goes across a contact interface, for example, the Al/TiN interface in a short stripe of Al on TiN, current crowding occurs near the cathode end as well as near the anode end in Figure 7.2a. The degree of current crowding depends not only on the resistance of Al and TiN but also on the contact resistance of the Al/TiN interface. The larger the contact resistance, the lesser the current crowding, as shown in Figure 7.2b by simulation, where the distribution of current density has spread widely. If the contact resistance is low, the distribution of current crowding tends to be narrower too. Furthermore, we note that the upper left corner has a very low current density, as given by the simulation of current distribution in the short stripe, as shown in Figure 7.2b.

The effect of current crowding on damage induced by electromigration is unusual because the damage tends to occur not in the region of high current density but in the region of low current density, which is against the expectation from the electron wind force, as discussed in Chapter 2. This is because in the low current density region, the electron wind force is low, so there should have been very little electromigration.

Today, the simulation of electric current distribution in 2D or 3D structures can be readily obtained because commercial programmes are available. In a simulation, we can easily see where the high current density regions are as well as where the low current density regions are. Under electromigration experiments, if a void forms in a high current density region, it is expected, but if it forms in a low current density region, an explanation is needed, which will be discussed below.

7.2 VOID FORMATION IN THE LOW CURRENT DENSITY REGION

Okabayashi et al. [8] prepared in-situ TEM samples, as depicted in Figure 7.3a, for a direct observation of void formation as well as hillock formation in an Al short stripe on TiN under electromigration. The width of the Al line was designed so that it could be imaged clearly in TEM from the side view rather than from the top view. In other words, the electron beam in the TEM is normal to the applied electron current of 50 mA from the

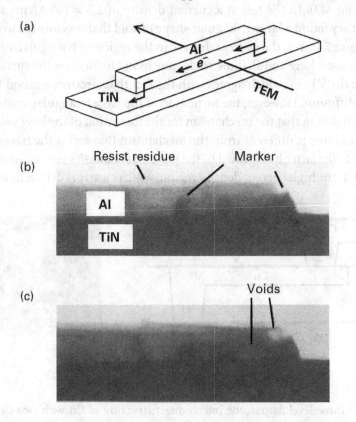

FIGURE 7.3   In-situ TEM samples and test results. (a) Schematic diagram of TEM test on the sample. (b) Void at the upper corner of the cathodic end. (c) Void formed at the upper tip of the hillock.

side view, as shown in Figure 7.3a. After 14 seconds of electromigration, a void was seen at the upper left corner of the stripe, i.e., the upper corner of the cathode end, as shown in Figure 7.3b. We note that the upper corner has a very low current density, as given by the simulation of current distribution in the short stripe, as shown in Figure 7.2b.

Furthermore, after a long time of electromigration, the polarity of the applied current was reversed, and a void was found to form at the upper tip of the hillock (which was the anode before the change of polarity), as shown in Figure 7.3c. Again, the current density at the upper tip of the hillock should be very low, yet the void there kept growing with electromigration. These results show that electromigration-induced void formation occurs in the low current density region.

Shingubara et al. [9] patterned U-shape baselines of TiN, on which Al short stripes were deposited. Some of the short stripes were made to stick out from the U-shape baseline of TiN, so that the Al short stripe has an overhang out of the baseline. The overhang was about 20 μm long. It is clear that there should be no current in the overhang during electromigration when current is applied to the TiN baseline. However, voids were observed experimentally to form in the overhang in electromigration.

Hu et al. [10] have conducted electromigration in a three-level damascene interconnect structure of Cu with two Cu vias (V1 and V2) connecting three levels of lines (M1, M2, and M3). In the test at a current density of 2.5×10^6 A/cm^2 at 295°C over 100 hours, they found a large triangular shape of void that was formed to the left of V1 via (see Figure 7.4). Yet, the current density in the region of triangular void formation is low. Also, some long and shallow voids were found to form on the surface of the line (M2) above the V1 via, which agrees with the fact that electromigration in Cu occurs by surface diffusion. However, the formation of the large triangular void is not.

It is worth noting that the mechanism for the formation of the long void on the surface of the M2 line is different from the mechanism that forms the triangular void to the left of V1. The latter has formed in the low current density region, while the former has formed in the high current density region, which clearly is driven by electron wind

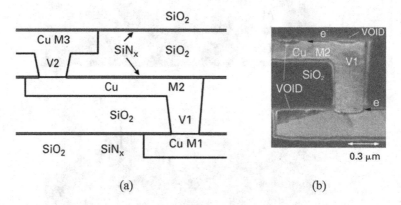

(a)   (b)

FIGURE 7.4   Three-level damascene interconnect structure of Cu with two Cu vias. (a) The cross-section view of the structure (b) A large triangular shape of void formed at the left of V1 via after electromigration test.

force. But what is the electromigration driving force to drive vacancies to go to the low current density region to form the triangular void? An explanation will be given later.

The observations presented above, showed that certain void formations induced by electromigration occur in the low current density regions in the Al and Cu interconnects of 3D IC. No doubt, this is quite unexpected on the basis of our understanding of the electron wind force of electromigration, which means that the lower the current density, the lower the driving force. Actually, in the low current density region, the electron wind force is so weak that there should be no electromigration damage. We further notice that all the phenomena of void formation in the low current density regions occurred near a turn of electron flow.

Owing to the fact that the electron wind force will keep driving more and more vacancy to the exit region, a void would have been nucleated there when the vacancy concentration in the exit region reaches the supersaturation needed for nucleation. The subsequent lateral growth of the void would have pushed the exit of electrons inward from the lower end because the void would now have occupied the lower end, so electrons have to move to the front of the void. Then, the lateral growth of the void would be fed by the ensuing flux of vacancies supplied by electromigration, so we expect that the void would have grown in a pancake-type shape along the Al/TiN interface (see Figure 7.2).

What is wrong with the above expectation is that it will not lead to the depletion of the entire cathode end of the Al short stripe. It only leads to the formation of a pancake-type void along the Al/TiN interface. However, this is not what has been observed. Actually, the void growth extends to the upper corner of the cathode [12]. Experimentally, the entire cathode of Al short stripe is depleted completely in electromigration.

Therefore, if we assume void formation starts from the cathode end at the lower left corner in Figure 7.2c, the lateral growth of the void will not be able to consume the upper part of the stripe, yet this is against the experimental observation that the entire cathode end of the stripe can be consumed.

In order to deplete the entire cathode, the vacancies must go to the low current density region at the upper left corner, as depicted in Figure 7.2. Therefore, the void must start or nucleate from the upper end of the left corner of the stripe, and the growth of the void should propagate downward as well as toward the anode. Consequently, a complete depletion of the cathode end can occur, and it will enable us to measure the depletion rate at the cathode end, which in turn enables us to measure Z^*.

We should examine whether or not the void nucleation at the upper left corner could be explained by stress migration or by thermomigration. If we assume that more and more vacancies are being driven to the high current density region (we could assume it to be in tension), a concentration gradient of vacancy is created between the high as well as the low current density regions. The latter can be assumed to have no stress. The gradient will drive vacancies from the high to the low current density region. Then the question is as follows: will it lead to void formation in the low current density region?

Because a vacancy concentration gradient has been assumed, the vacancy concentration in the high current density region is always higher than that in the low current density region. Yet, the nucleation of a void requires the supersaturation of vacancies. Then, because the vacancy concentration in the high current density region is higher, it is unreasonable to assume that voids will be nucleated in the low current density region rather than in the higher current density region.

For thermomigration, it will be helpful if we review briefly what will happen to a Cu kettle when it is used to boil water. The outer temperature of the kettle will be over 600°C, and the inside will be about 100°C. If the thickness of the wall of the kettle is 1 mm, we have a temperature gradient of 5,000°C/cm, which is very large [7]. Therefore, we expect thermomigration to occur across the wall of the kettle. The temperature gradient will drive Cu atoms to diffuse from hot to cold, from the outside to the inside. If it had happened, we should have expected the inside diameter of the kettle to enlarge with time because more and more atoms diffused into it. Somehow, if we examine the kettles at home after years of use, they do not become larger at all.

The reason is that at the hot end, there are more vacancies, and the diffusion of vacancies tends to balance the thermomigration of atoms. Thus, thermomigration in a pure metal tends to be negligible. Indeed, we did not consider thermomigration in Cu and Al interconnects. However, only when solder joint technology has reliability problems do we have to consider thermomigration. The topic will be covered in Chapter 8.

7.3 CURRENT DENSITY GRADIENT FORCE IN ELECTROMIGRATION

Generally speaking, the driving force of electromigration is due to electron wind force; however, it cannot explain void formation in the low current density region, especially in 3D IC technology. Therefore, a new driving force of electromigration, the "current density gradient force," which is normal to the electron wind force, has been proposed [10,11,12].

The new driving force will divert vacancies to diffuse to the low current density region before they reach the high current density region driven by the electron wind force. Because vacancies go to the low current density region, the concentration of vacancies can reach supersaturation, and voids can nucleate there. However, it is an uphill diffusion of vacancies.

In Figure 7.5a, we consider the electromigration in a conductor that has a 90° turn. The proposed "current density gradient force" comes from the existence of a gradient of electric potential in the region of current crowding due to the turn. Figure 7.5b shows the simulation of the electrical potential gradient in the region of the turn.

We assume that the conductor is straight before the turn, so the current density is uniform in the straight conductor. At the same time, a uniform vacancy flux moves from the anode side to the cathode side, or an atomic flux moves in the opposite direction. Approaching the turn, current crowding occurs, so the current density is higher on the inner side of the turn and lower on the outer side of the turn. For those

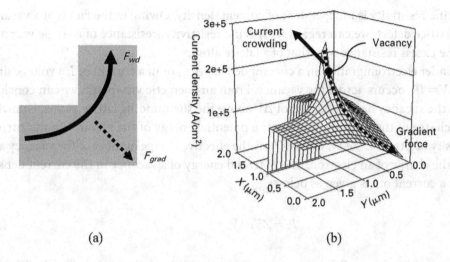

FIGURE 7.5 The void formation in the low current density region. (a) the schematic diagram of current density gradient force and electron wind force. (b) The simulation result of electrical potential gradient in the region of the turn.

vacancies that follow the high current density path on the inner side, their potential energy increases; therefore their equilibrium concentration should decrease with respect to that in the uniform region before the turn.

However, the low current density region on the outer side of the turn can have a higher equilibrium vacancy concentration than that in the uniform region. Thus, the electric potential gradient will push the excess vacancies in the high current density region to the low current density region. In other words, the force of the electric potential energy gradient drives vacancies to go from the high current density region to the low current density region.

Clearly, the diffusion of vacancies is uphill, going from a low vacancy concentration region to a high vacancy concentration region, i.e., against the vacancy concentration gradient. However, the uphill flux of vacancy is actually driven by the downhill electric potential energy gradient. Quantitatively, an analysis is given below.

When an electric potential is applied to the conductor, the chemical potential energy of every atom and vacancy is increased. Because of resistance, the increase in potential energy of vacancies (or solute atoms) is much larger than that of lattice atoms. Furthermore, the increase is proportional to the increase in current density. In the high current density region, the vacancies will have a higher potential energy, so the concentration of vacancies in the high current density region will be reduced. In other words, it means that the formation energy of a vacancy in the high current density region is higher, so the concentration will be lower, similar to that in a mechanically compressed solid.

Now, we consider a single crystal Al stripe and assume that a vacancy in the Al crystalline lattice has a specific resistivity of $\rho_v = R_v$. The specific resistivity may depend on current density due to Joule heating since resistivity depends on temperature. However, for simplicity, we ignore the Joule heating effect here and assume that the specific resistivity is independent of current density. Owing to the fact that a vacancy is a lattice defect, we can regard its specific resistivity (resistance of a single vacancy) as the excess resistance over that of a lattice atom.

Under electromigration in a current density of j_e, or in a current of I, a voltage drop of $\Delta V = IR_v$ occurs across the vacancy. From an energetic viewpoint, we can conclude that the vacancy has a potential of ΔV above the surrounding lattice atoms. Knowing the charge of the vacancy, we have the potential energy of the vacancy in the current density of j_e to be $Z^{**}e\Delta V$, where Z^{**} is the effective charge number of the vacancy and e is the charge of an electron. The potential energy of a vacancy in the current density j_e or a current of I is given as below:

$$P_v = Z^{**}e\Delta V \tag{7.1}$$

If we assume the equilibrium vacancy concentration in the single crystal with zero electrical current ($j_e = 0$) to be C_v,

$$C_v = C_0 \exp(-\Delta G_f / kT) \tag{7.2}$$

where C_0 is the atomic concentration of the crystal and ΔG_f is the formation energy of a vacancy in the crystal. When we apply a current density of j_e to the crystal, the vacancy concentration will be reduced to

$$C_{ve} = C_0 \exp-(\Delta G_f + Z^{**}e\Delta V)/kT \tag{7.3}$$

Under a uniformly high current density, the equilibrium vacancy concentration in the crystal decreases. In other words, the electric current dislikes any excess high-resistive obstacles (or defects) and prefers to get rid of them until equilibrium is reached. When there is a current density gradient as in current crowding, a driving force exists to do so,

$$F = -dP_v / dx \tag{7.4}$$

where P_v has been defined in Eq. (7.1) as the electric potential energy of a vacancy in a conductor having the current density j_e. This force drives the excess vacancies to diffuse in the direction normal to the direction of electron flow. Consequently, a component of the vacancy flux is moving in the direction normal to the electron flow, as shown in Figure 7.5b,

$$J_{cc} = C_{ve}(D_v / kT)(-dP_v / dx) \tag{7.5}$$

where D_v/kT is the mobility and D_v is the diffusivity of vacancies in the crystal.

Because a constant flux of vacancies keeps coming from the anode to the cathode due to electromigration, the total flux of vacancies moving towards the cathode is given by the vector sum of the two components below,

$$J_{sum} = J_{em} + J_{cc} = C_{ve}\left(D_v/kt\right)\left(-Z^*eE - dP_v/dx\right) \tag{7.6}$$

where the first term, J_{em}, is due to electromigration driven by the current density (the electron wind force) and the second term, J_{cc}, is due to current crowding, driven by the current density gradient force.

In the first term, Z^* is the effective charge number of the diffusing Al atom, and $E = j_e\rho$ (where ρ is the resistivity of the Al). Here, we note that we assume the vacancy flux is opposite but equal to the Al flux. Also, it is important to note that the sum in the bracket in Eq. (7.6) is a vector sum; the first term is directed along the electron current direction, and the second term is directed normal to the electron current direction. Therefore, the vacancies are driven by two forces in the current crowding region. They are depicted in Figure 7.5a in the upper left corner. Because the current turns continuously in the current crowding region, the direction of J_{sum} changes with position.

How large is the gradient force? If we apply a current density of 10^5 A/cm² through the stripe and assume that the current density will drop to zero across the thickness of the stripe of 1 μm, the gradient can be as high as 10^9 A/cm³. The gradient force is of the same magnitude as the electron wind force.

However, the electric potential gradient across an atom is quite small, so the current density gradient force on a lattice atom is negligible. Nevertheless, because the electrical resistance of a vacancy is about one hundred times greater than that of an atom, the electric potential gradient force on a vacancy is significant.

From the kinetic point of view, the diffusion of vacancies driven by the current density gradient force occurs mainly by lattice diffusion. While it is possible to consider grain boundary diffusion if a grain boundary happens to exist in the region of current crowding, it is hard to imagine how it can occur by surface diffusion because there is no surface diffusion path in the turn. Nevertheless, vacancy diffusion and interfacial diffusion can occur in Cu at the device operation temperature of 100°C.

In a short summary, we postulate here that defects such as vacancies and solute atoms have a higher potential energy in the high current density region than that in the low current density region. The potential energy gradient in the current crowding area provides a new driving force to push these defects to diffuse from the high current density region to the low current density region. As a consequence, the voids tend to form in the low current density region rather than in the high current density region. Furthermore, electromigration failure tends to occur in a low current density region in 3D IC interconnects because of current crowding.

7.4 CURRENT CROWDING-INDUCED PANCAKE-TYPE VOID FORMATION IN FLIP-CHIP SOLDER JOINTS

In flip-chip solder joint technology, the solder bump has a diameter of about 100 μm and is connected to an Al or Cu line of 1 μm or so in thickness. If we assume the width of the line and the diameter of the bump are the same, there is a change in current density by two orders of magnitude when the same current is passing through them. A very large current crowding occurs in the line-to-bump contact or in the transition region of a thin under-bump metallization (UBM). Combining atomic flux divergence and current crowding, the thin UBM contacts have been the most important failure site of electromigration in flip-chip solder joints, discussed in Chapter 4.

To overcome the failure due to pancake-type void formation, a thick UBM decreases current crowding. Simulation showed that when a thick UBM is used, over 10 μm of Cu, the current density distribution will spread over the entire contact, resulting in little current crowding. For this reason, a pancake-type void will not form; instead, a group of smaller voids will form at the interface.

7.5 CURRENT CROWDING-INDUCED PASSIVATION OPENING IN CU-TO-CU DIRECT BONDING

In Chapter 4, Section 4.8, we have discussed pancake-type void formation at the passivation opening in Cu-to-Cu direct bonding. At the entrance of the passivation opening, current crowding occurs, which leads to a similar type of pancake-type void formation as in flip-chip solder joints.

On the basis of the discussions presented in this section and in the last section, it is clear that when current crowding occurs in any contact opening where there is an interface nearby, pancake-type void formation can occur. This is a general mechanism of failure induced by electromigration at thin contact interfaces. No doubt, a thick UBM will overcome this kind of failure due to current crowding. However, in micro-bump technology, where a thick Cu UBM has been used, it has introduced another type of failure due to porous IMC formation.

PROBLEMS

7.1. Electric current crowding occurs in the multi-layer interconnect structure when there is a turn in the path of conduction. Besides turns, what are other structural features that can lead to current crowding?

7.2. In flip-chip solder joints, it is known that a thin UBM can lead to pancake-type void formation due to current crowding. Thick UBM has been used to avoid it. Yet, another kind of failure mode of porous IMC formation can occur! How can we avoid porous IMC formation?

7.3. While we have current crowding, do we have stress crowding or temperature crowding?

7.4. In TSV structure, do we consider current crowding? Why not?

7.5. What is the effect of current crowding on Joule heating?

REFERENCES

[1] E. C. C. Yeh, W. J. Choi, K. N. Tu, P. Elenius, and H. Balkan, "Current-crowding-induced electromigration failure in flip chip solder joints," *Appl. Phys. Lett.*, 80(4), 580–582 (2002).

[2] L. Zhang, S. Ou, J. Huang, K. N. Tu, S. Gee, and L. Nguyen, "Effect of current crowding on void propagation at the interface between intermetallic compound and solder in flip chip solder joints," *Appl. Phys. Lett.*, 88, 012106 (2006).

[3] M. Li, D. W. Kim, S. Gu, D. Y. Parkinson, H. Barnard, and K. N. Tu, "Joule heating induced thermomigration failure in un-powered micro-bumps due to thermal crosstalk in 2.5D IC technology," *J. Appl. Phys.*, 120, 075105 (2016).

[4] K. N. Tu, "Electronic thin film reliability," in *Irreversible Processes in Interconnect and Packaging Technology*, Cambridge University Press, Cambridge, UK (2011) (Chapter 10).

[5] C. C. Yeh and K. N. Tu, "Numerical simulation of current crowding phenomena and their effects on electromigration in VLSI interconnects," *J. Appl. Phys.*, 88, 5680–5686 (2000).

[6] K. N. Tu, C. C. Yeh, C. Y. Liu, and C. Chen, "Effect of current crowding on vacancy diffusion and void formation in electromigration," *Appl. Phys. Lett.*, 76, 988–990 (2000).

[7] V. S. Arpact, *Conduction Heat Transfer*, Addition-Wiley, Reading, MA (1966), p. 44.

[8] H. Okabayashi, H. Kitamura, M. Komatsu, and H. Mori, "*In-situ* side-view observation of electromigration in layered Al lines by ultrahigh voltage transmission electron microscopy," *AIP Conf. Proc.*, 373, 214 (1996).

[9] S. Shingubara, T. Osaka, S. Abdeslam, H. Sakue, and T. Takahagi, "Void formation mechanism at no current stressed area," *AIP Conf. Proc.*, 418, 159 (1998).

[10] C. K. Hu, L. Gignac, S. G. Malhotra, R. Rosenberg, and S. Boettcher, "Mechanisms for very long electromigration lifetime in dual-damascene Cu interconnects," *Appl. Phys. Lett.*, 78, 904 (2001).

[11] F.-Y. Ouyang, K. Chen, K. N. Tu, and Y.-S. Lai, "Effect of current crowding on whisker growth at the anode in flip chip solder joints," *Appl. Phys. Lett.*, 91, 231919 (2007).

[12] H. Ye, C. Basaran, and D. C. Hopkins, "Thermomigration in Pb-Sn solder joints under joule heating during electric current stressing," *Appl. Phys. Lett.*, 82, 1045–1047 (2003).

CHAPTER 8

Effect of Joule Heating on Electromigration

8.1 INTRODUCTION

In consumer electronic products, the continuing demand for smaller size, more functionality, and lower power consumption is challenging in both manufacturing and failure prediction. At the moment, the two major trends in electronic packaging technology are the need for a higher and higher number of input/output (I/O) counts for better signal resolution and the demand of lower and lower power consumption in consumer electronic products.

Low power means low Joule heating, as discussed in Chapter 5, Section 5.3. However, a higher number of I/O means a smaller diameter in flip-chip solder joints or in Cu-to-Cu direct bonding, but either one will increase Joule heating.

In dense 2.5D IC or 3D IC devices, the addition of a Si interposer requires the addition of one more redistribution layer (RDL) and one more level of solder joints. Consequently, not only Joule heating is serious, but also heat dissipation is poor. To enhance heat dissipation, we need to have a high temperature gradient. Unfortunately, the high temperature gradient can lead to thermomigration, to be discussed below.

In electronic packaging microstructure, just 1°C across a solder joint of 10 μm in diameter is equal to 1,000°C/cm, which is a very large temperature gradient from the point of view of thermomigration. Furthermore, the use of Si interposer enhances the lateral heat transfer along the interposer, which has been found to have caused an unexpected thermomigration failure (see Figure 8.1).

Therefore, thermal management affects nearly all the reliability problems in mobile technology. Electromigration enhanced by Joule heating is the primary concern for failure, especially early failure. In this chapter, we shall discuss the synergistic effect between electromigration and Joule heating and thermal crosstalk on a Si interposer.

DOI: 10.1201/9781003384281-8

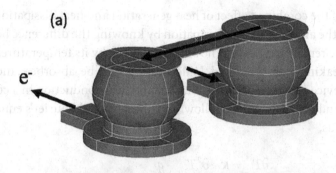

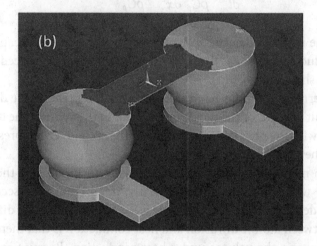

FIGURE 8.1 Simulation of current density in solder joints during the electromigration test. (a) Mesh and model of a pair of solder joints. (b) The simulation result of current density distribution in a pair of solder joints.

8.2 HEAT GENERATION AND HEAT DISSIPATION

8.2.1 Joule Heating Measurement

Heat generation in electronic devices is mainly due to electrical resistance, or Joule heating, which is entropy production on the basis of irreversible processes, as presented in Chapter 5. For electrical conduction, we have shown that entropy production is the product of the conjugated flux of j (current density=coulomb/cm²-sec) and the conjugated driving force of E (electric field $E=j\rho$, where ρ is resistivity).

$$\frac{TdS}{Vdt} = jE = j^2\rho \tag{8.1}$$

where T is temperature, V is volume of sample, dS/dt is entropy production rate, and $j^2\rho$ is Joule heating per unit volume per unit time, in units of Watt/cm³. This means that by measuring the flux and the driving force, we can calculate the Joule heating generated by electrical conduction.

However, heat dissipation is harder to measure. On the other hand, we can measure the temperature of the sample. It is worth noting that the sample temperature is the consequence of the combined effect of heat generation and heat dissipation. Therefore, we can obtain the amount of heat dissipation by knowing the difference between Joule heating and the remaining heat in the sample as given by its temperature.

General speaking, the generated Joule heating will be absorbed and conducted through the device structure. Heat generation and heat conduction in a conductor are fully coupled and governed by the following 1-dimensional Fourier's equation of heat transfer in solids,

$$\frac{\partial T}{\partial t} = \frac{\kappa}{\rho C_p} \frac{\partial^2 T}{\partial x^2} + \frac{q}{\rho C_p} \tag{8.2}$$

where ρ is the material density of the conductor, C_p is the heat capacitance, κ is the thermal conductivity, and q is the power density (Watt/cm³) induced by Joule heating, and $q = jE$, as shown in Eq. (8.1).

The first term on the right-hand side of Eq. (8.2) represents the divergence of heat fluxes in a unit volume, or the net change of heat in the volume due to heat flux in and heat flux out with the neighbouring volumes. The second term represents heat generation within the unit volume due to Joule heating.

It is worth mentioning that the unit of $(\kappa/\rho C_p)$ in Eq. (8.2) is the same as atomic diffusivity D with a unit of cm²/sec. Clearly, if we ignore the second term on the right-hand side of Eq. (8.2), it is the same as Fick's second law of diffusion. Thus, the difference between atomic diffusion and heat conduction is that there is always a temperature change in the volume under heat conduction. Because entropy is positive, heat is generated in the volume, so the temperature of the volume increases.

8.2.2 Heat Dissipation Measurement

Dissipation of Joule heating into the ambient occurs in complex ways, depending on the device structure and materials as well as the ambient. To simplify the picture, we consider a steady-state model in which the heat is dissipated in terms of heat flux to the ambient in two ways. The first is through Si to the surrounding ambient, and the second is through the substrate to the surrounding ambient. Typically, heat flux, according to Fourier's law, is given as $J_{heat} = -\kappa \, (dT/dx)$, which is a vector. Here, we assume experimentally that it is governed by the following equation,

$$J_{heat} = (h_{Si} + h_{sub})(T - T_{ext}) \tag{8.3}$$

where J_{heat} is the heat dissipation by heat flux (joule/cm²-sec or Watt/cm²), h is the experimental thermal transfer coefficient (W/(cm²·K)). T_{ext} is the ambient temperature and T is the temperature of the conductor under consideration. For Si, we may assume h_{Si} to be 900 W/(cm²·K), and for the substrate, we may assume h_{sub} to be 5 W/(cm²·K). With these assumed values of the thermal transfer coefficient, it is clear that most heat is dissipated through the Si side of the model. Models of heat transfer simulation are readily available.

To increase the area of the Si surface in order to enhance the heat dissipation, a fin structure can be attached to the Si surface or the substrate surface by using the thermal interface material (TIM), which should have an excellent thermal transfer coefficient. For this reason, the search for TIM has been active.

8.2.3 Temperature Measurements

As we have mentioned before, in 3D IC, a very large temperature gradient can exist and can lead to thermomigration failure. For example, a temperature difference of 5°C across a μ-bump of 20 μm in diameter will produce a temperature gradient of 2,500°C/cm. Therefore, we need to measure temperature accurately, within ±1°C. Below, we shall discuss two of the common ways to measure temperature: the first is by infrared (IR), and the second is by electrical resistance.

Figure 8.2 shows the IR measurement of temperature distribution on the cross-sectional surface of a flip-chip solder joint with a diameter of about 100 μm, with

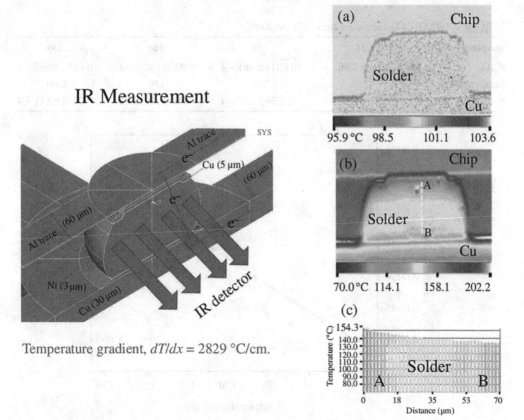

FIGURE 8.2   Infar-red (IR) measurement of temperture gradeint across a flip chip solder joint (a) Before joule heating, (b) With Joule heatting, and (c) Temerature distribution across the solder joint.

an Al trace on top and a Cu trace at the bottom. Figure 8.2a and b on the right-hand side shows, respectively, the IR measurement before and after applying a high current density. Figure 8.2c shows the temperature distribution across the solder joint. The temperature gradient measured in Figure 8.2b is about 2,829°C/cm.

To measure the temperature in a set of solder balls by electrical resistance, we need to determine the resistance of a solder ball as a function of temperature, or what is called the temperature coefficient of resistance (TCR) test. Table 8.1 shows the resistance, $R_N(\Omega)$, as a function of temperature from 25°C to 200°C. The data are plotted on the curve as shown in Figure 8.3. From the curve, we obtain the slope, tan $\Theta = dR/dT = 1.42 \times 10^{-5}$ Ω/°C. For example, if we consider the temperatures of 120°C and 60°C, the resistance values are 0.00525 and 0.0045 Ω, respectively. Now, we define tan $\Theta = \Delta R/\Delta T = 1.42 \times 10^{-5}$ Ω/°C to be the TCR.

To determine the increase in temperature due to Joule heating, the Joule-heated sample resistance (R_J) is obtained after applying the test electrical current. The temperature rise due to Joule heating is obtained from the following relationship:

$$\Delta T = \frac{R_J - R_N}{dR/dT}$$

TABLE 8.1 $R_N(\Omega)$ of Temperature from 25°C to 200°C

Temperature (°C)	25	50	100	130
$R_N(\Omega)$	0.0041±1.0E−4	0.0043±1.0E−4	0.0050±1.7E−4	0.0054±2.6E−4
Temperature (°C)	150	160	170	200
$R_N(\Omega)$	0.0057±2.6E−4	0.0059±2.3E−4	0.0060±2.3E−4	0.0066±2.1E−4

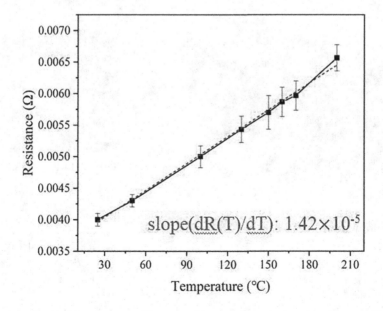

FIGURE 8.3   Resistance measurement of temperature calibration.

TABLE 8.2   The Electromigration Conditions

Oven Temp. (°C)	Voltage (mV)	Current (A)	Current Density (A/cm²)	R_J (Ω)	Δ (°C)
160	85.5	9	4,586	0.0095	210

where R_N is the non-Joule-heated sample resistance at the oven temperature, which is equal to the EM test temperature, provided that the oven temperature is uniform.

To calculate ΔT, we take R_N=0.0059 Ω, R_J=0.0080 Ω, and dR/dT=1.42 × 10⁻⁵ Ω/°C. We obtain ΔT=210°C. Because the oven temperature is 160°C, so the temperature increase due to Joule heating is 210°C − 160°C = 50°C. The electromigration conditions are given in Table 8.2.

8.3 JOULE HEATING AND ELECTROMIGRATION

How to couple Joule heating and electromigration requires careful consideration because we may have heat flow, charge flow, and atomic flow together. It may mean that we have a 3 × 3 matrix equation. To simplify the case, we consider below the physical picture of the coupling or no coupling between Joule heating and electromigration.

Joule heating will increase the temperature of the sample, but at the same time, dissipation of heat will create a temperature gradient. If the temperature gradient is large, it may induce thermomigration. Both the temperature increase and the temperature gradient affect atomic diffusion in electromigration. First, we consider no interaction between Joule heating and electromigration. Then, we consider thermomigration.

8.3.1 No Effect of Joule Heating on Electromigration in Al Short Stripes below the Critical Length

In Chapter 3, Figure 3.3 shows a set of Al short stripes deposited on a TiN line for electromigration study. It was found that below the critical length, there was no electromigration damage in the short stripe, as shown by the three short stripes near the upper right corner of the figure.

This phenomenon was explained by Blech and Herring due to the back-stress effect. However, we note that while there is no electromigration damage, electrical conduction still occurs in those short stripes, so there is Joule heating in them. Clearly, this finding has decoupled the link between Joule heating and electromigration. Nevertheless, electromigration occurs through atomic diffusion, which can be affected by temperature increases. Therefore, the effect of Joule heating on electromigration is mainly due to temperature increases.

8.4 JOULE HEATING-INDUCED INTERACTION BETWEEN THERMOMIGRATION AND ELECTROMIGRATION

In Figure 4.3a, a schematic diagram depicts the cross-sectional view of a pair of flip-chip solder joints. The top part of the joints is connected by an Al line, and the bottom parts are connected to Cu pads. In Figure 4.3b, the dotted arrows indicate the flow direction of electrons. The pair of blue arrows indicate the direction of electromigration due to the flow of atoms, which also follows the direction of the dotted arrows. The blue arrow on the left-hand side is directed upward, but the one on the right-hand side is directed downward.

The pair of red arrows in Figure 4.3b indicate the direction of thermomigration. It is worth emphasizing that the temperature gradient due to the large temperature difference across the solder joint is because of the large resistance difference between the Al line above the solder joint and the Cu pads below the solder joint. On the other hand, if we only consider electromigration along a Cu or Al interconnect line, the Joule heating of electrical conduction may induce a small increase in temperature but not a large increase in temperature gradient.

In Figure 4.3b, because the Joule heating in the Al line is much higher than that in the Cu pads, there is a large temperature gradient across each of the solder joints, so the red arrows indicate thermomigration from top to bottom. It is worth noting that in the solder joint on the left-hand side, the direction of electromigration is opposite that of thermomigration. But in the solder joint on the right-hand side, they are in the same direction.

To combine thermomigration and electromigration, we could add or subtract the corresponding fluxes, as shown below:

$$J = C \frac{D}{kT} \frac{Q^*}{T} \left(-\frac{\partial T}{\partial x} \right) \pm C \frac{D}{kT} Z^* e \rho j \qquad (8.4)$$

However, Eq. (8.4) could be misleading because we must recall the classical finding of the Sorel effect on thermomigration. Under a temperature gradient, a homogeneous alloy becomes inhomogeneous, in which one of the components moves from the hot end to the cold end while the other component moves from the cold end to the hot end. Therefore, we need to define the flux of which of the components in Eq. (8.4).

Let us consider only thermomigration below:

$$J = C \frac{D}{kT} \frac{Q^*}{T} \left(-\frac{\partial T}{\partial x} \right) \qquad (8.5)$$

In Eq. (8.5), the heat of transport, Q^* can be positive or negative. To define the sign of Q^*, we examine Eq. (8.5). We shall consider J in the equation between two points: point 1 at $(x1, T1)$ and point 2 at $(x2, T2)$ in a Cartesian coordinate, as shown in Figure 8.4, and we assume $T_1 > T_2$, and atomic flux moves from hot to cold, i.e., from point 1 to point 2. Then, $\Delta T/\Delta x$ is negative, so Q^* is positive. Recall that this is also the reason why all the flux equations shown in Chapter 5 have a negative sign. Thus, for an element that moves from the hot end to the cold end, Q^* is positive. For an element moving from cold to hot, its Q^* is negative [1–4].

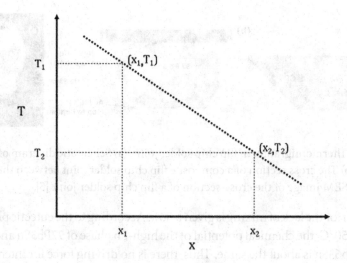

FIGURE 8.4 The two points used to determine the sign of Q^*.

Therefore, in Figure 4.3b, the red arrows apply to the thermomigration of the component, which moves from the hot end to the cold end. For the other component, which moves from the cold end to the hot end, we need to change the direction of the red arrows.

In Section 8.4.1, we discuss thermomigration in composite solder joints because we can observe thermomigration clearly.

8.4.1 Thermomigration in Flip-Chip Composite Solder Joints

In Figure 8.5, (a) shows the schematic diagram of a flip chip on a substrate, (b) shows the cross-section of a composite flip-chip solder joint between the chip and the substrate, and (c) shows an SEM image of the cross-section of a composite flip-chip solder joint.

In Figure 8.5 (a), the small squares on the substrate are electrical contact pads. The composite solder joint was composed of high-Pb solder of 97Pb3Sn on the chip side and eutectic 37Pb63Sn on the substrate side. The solder bump has a height of about 100 μm. The contact opening on the chip side has a diameter of 90 μm. The tri-layer thin films of under-bump-metallization (UBM) on the chip side were Al (~0.3 μm), Ni(V) (~0.3 μm), and Cu (~0.7 μm). On the substrate side, the bond-pad metal layers were Ni (5 μm) coated with a thin film of Au (0.05 μm).

As a control experiment, constant temperature annealing of the composite flip-chip samples was performed in an oven at 150°C in ambient conditions for a period of 1 week, 2 weeks, and 1 month. The microstructures of the cross-section of the composite solder joint were examined under an optical microscope and a scanning electron microscope, as shown in Figure 8.5b. The composition was analysed using energy dispersive X-ray (EDX) and electron probe microanalysis (EPMA). However, no mixing or homogenization between the high-Pb and the eutectic was observed after the 1-month annealing because the images were nearly the same as those shown in Figure 8.5c.

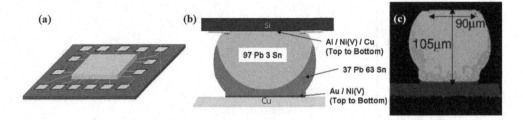

FIGURE 8.5   Thermomigration in flip chip solder joint. (a) Schematic diagram of a flip chip on a substrate, (b) The cross section of a composite flip chip solder joint between the chip and the substrate, (c) SEM image of the cross-section of a flip chip solder joint [5].

The reason for the lack of mixing is given below. According to the eutectic phase diagram of Sn-Pb, at 150°C, the chemical potential of the high-Pb phase of 97Pb3Sn and the eutectic phase of 37Pb63Sn is about the same. Thus, there is no driving force for intermixing.

To detect thermomigration by using the temperature gradient induced by Joule heating, a set of 24 composite solder bumps on the peripheral of the Si chip was tested. It was discussed in Chapter 5, Section 5.7. Thus, it will not be repeated here.

8.4.2 Effect of Temperature Increase on Electromigration and Vice Versa

Here, we consider the effect of a uniform temperature increase due to Joule heating on electromigration. It means the rate of electromigration will be faster at a high temperature due to faster atomic diffusion. However, in order to simplify the case, we assume the temperature increase (ΔT) due to Joule heating is relatively small, so it will not cause thermomigration. Therefore, we just need to consider the effect of ($T+\Delta T$) on electromigration.

In the opposite direction, about the effect of electromigration on Joule heating, when electromigration leads to a change in resistance or a change in the current density of the sample, Joule heating will increase. To increase current density means to reduce the cross-section of the sample, assuming that the applied current remains unchanged. In Cu interconnects, electromigration occurs by surface diffusion on the Cu surface near the cathode end, and it tends to reduce the thickness as well as the cross-section of the sample, as shown in Figure 4.2. When the cathode end is becoming thinner and thinner, the current density should increase, so Joule heating as well as the temperature will increase. In turn, it will increase the rate of electromigration. This positive feedback will accelerate failure quickly, as discussed in Chapter 4.

8.5 JOULE HEATING-ENHANCED ELECTROMIGRATION FAILURE OF RDL

In a simple 2.5D IC device, there are two pieces of Si chips, one on top of the other, and there are three levels of solder joints. To interconnect them, two sets of RDL are needed due to circuit fan-out in order to increase the number of I/O counts. When the design of RDL is not given enough reliability consideration, it becomes the weak link in electromigration. It was caused by Joule heating and the positive feedback between Joule heating and electromigration to enhance electromigration failure. In Chapter 4, the dramatic failure mode of melting of the RDL was shown, so it will not be repeated here.

8.6 JOULE HEATING-INDUCED THERMOMIGRATION DUE TO THERMAL CROSSTALK

Thermal-crosstalk-induced thermomigration failure in un-powered micro-bumps has been found in 2.5D IC devices. In the test structure, as shown in Figure 8.6, there are two pieces of Si chips placed horizontally on a Si interposer. Figure 8.6a and b show, respectively, an SEM cross-sectional image and an X-ray tomography image of the top view of the test sample. In Figure 8.6a, a stacking of two levels of Si chips is shown; the bottom larger Si chip serves as an interposer, and it is connected vertically to the top two smaller Si chips through thermal compression bonding of micro-bumps, which are arranged in daisy chains, as shown in Figure 8.6b. The top two Si chips are separated by underfill materials. Under each of these two chips, there are two daisy chains of micro-bumps measuring 17 μm in diameter and 30 μm in pitch. The design of the daisy chain circuit of the micro-bumps is shown in Figure 8.6c.

One daisy chain of the micro-bumps under one Si chip (Chip 2) in Figure 8.6b was powered at a current density of 5.3×10^4 A/cm². The temperatures used for testing were 150°C and 170°C. Owing to Joule heating in the daisy chain, the temperature in the powered micro-bumps will rise about 30°C (estimated by simulation) above the test temperature. Yet there is little temperature gradient in the powered micro-bumps under Chip 2 because of the daisy chair arrangement. So, the powered micro-bumps have electromigration but no thermomigration.

However, the un-powered micro-bumps in the neighbouring chip (chip 1) in Figure 8.6a were damaged with big holes in the solder layer. This is because the Joule heating from the powered micro-bumps was transferred horizontally, through the

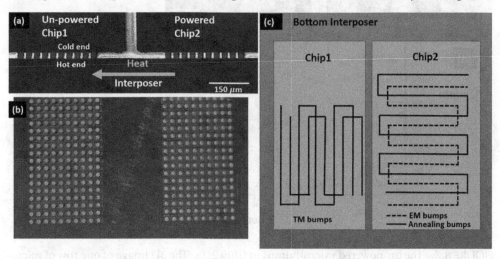

FIGURE 8.6 The damaged un-powered micro-bumps (a) SEM cross-sectional image of the test sample, (b) X-ray tomography image of the top view of the test sample. (c) The design of the daisy chain circuit of the micro-bumps.

interposer, to the bottom of the neighbouring un-powered micro-bumps, and created a large temperature gradient, in the order of 1,000°C/cm, across the un-powered micro-bumps in the neighbouring chip 1, so the latter bumps failed by thermomigration.

Furthermore, as one of the daisy chains in Chip 2 was powered, the Joule heating will be transferred to the un-powered micro-bumps in the neighbouring chain in Chip 2, through both the top Si chip and the bottom Si interposer. Therefore, the un-powered micro-bumps in Chip 2 have no temperature gradient, but they will experience a higher temperature annealing, which is slightly higher than the ambient temperature by 30°C due to Joule heating, but neither electromigration nor thermomigration.

Figure 8.7a shows the 3D image of one row of micro-bumps in the as-received state by using synchrotron radiation X-ray micro-tomography. Figure 8.7b shows the

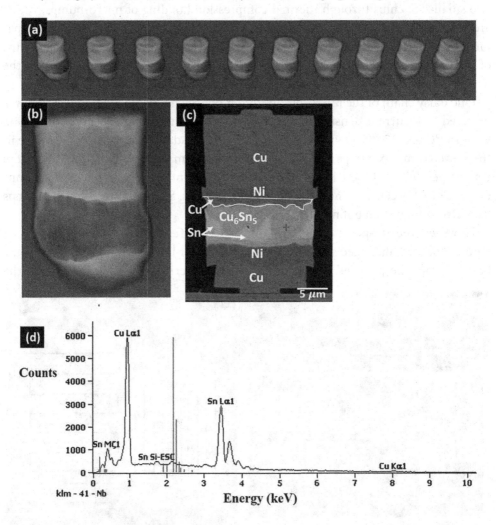

FIGURE 8.7 The un-powered micro-bumps in Chip 2. (a) The 3D image of one row of micro-bumps in the as-received state by using synchrotron radiation x-ray micro-tomography. (b) Synchrotron radiation image of one micro-bump. (c) The composition of a micro-bump. (d) X-ray diffraction pattern of the red cross-point in the micro-bump.

synchrotron radiation image of one micro-bump. Figure 8.7c shows the composition of a micro-bump. Figure 8.7d shows the X-ray diffraction pattern of the red cross-point in the micro-bump.

Figure 8.8 shows the 3D image of an array of micro-bumps after failure induced by thermomigration.

Figure 8.9a shows the odd row (row 1) micro-bumps under Chip 2, under iso-thermal annealing. Figure 8.9b shows the micro-bumps under Chip 2, which were stressed by electromigration at 120 mA and 170°C for 72 hours. Figure 8.9c shows the micro-bumps in Chip 1 that were under thermomigration. In comparison, those micro-bumps in Chip 1 under thermomigration have had more damage than the others.

In short summary, there are three different sets of micro-bumps in the test sample: (a) the micro-bumps under electromigration (the powered micro-bumps in Chip 2); (b) the micro-bumps under thermomigration (the micro-bumps in Chip 1); and (c)

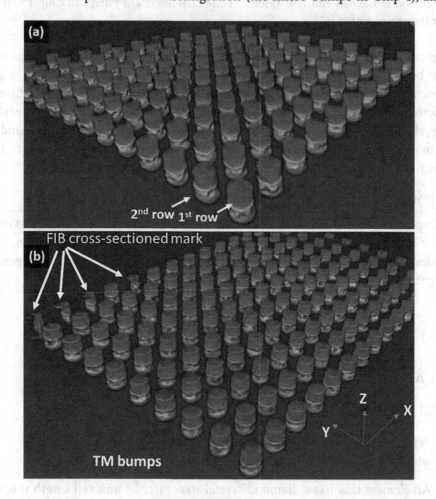

FIGURE 8.8   the 3D image of an array of micro-bumps after failure induced by thermomigration.

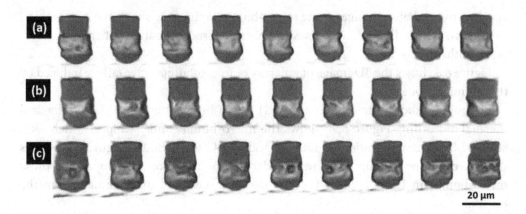

FIGURE 8.9 The micro-bumps in Chip 1. (a) The odd row (row 1) micro-bumps under Chip 2 under iso-thermal annealing. (b) The micro-bumps under Chip 2 that were stressed by electromigration at 120 mA and 170 °C for 72 hours. (c) The micro-bumps in Chip 1 that were under thermomigration [6].

the micro-bumps under a higher temperature isothermal annealing (the un-powered micro-bumps in Chip 2).

Synchrotron radiation tomography was used to inspect and compare these three sets of micro-bumps, as shown in Figure 8.9. In order to check the damage inside the micro-bump, the samples were analysed slice by slice in the Avizo software. It was found that even in the middle of the micro-bump, those under thermomigration had at least twice as much void volume as those under electromigration micro-bumps [6–7].

8.7 JOULE HEATING ON MEAN-TIME-TO-FAILURE EQUATION

In Chapter 9, a formal derivation of mean-time-to-failure (MTTF) equations on the basis of entropy production will be given. We present here the MTTF equation, which includes the effect of Joule heating on electromigration by modifying the temperature change of ΔT.

$$t^{failure} = MTTF = A\left(\frac{T+\Delta T}{j}\right)^2 \exp\left(\frac{E}{k(T+\Delta T)}\right)$$

where ΔT is the temperature increase due to Joule heating.

PROBLEMS

8.1. What is Joule heating? Please derive that the power of joule heating is $P = I^2 R$, where I is current and R is resistance.

8.2. An element that has a diamond crystal structure. Its unit cell length is 0.5 nm. What is the atomic volume of this element?

8.3. In a flip-chip solder joint, the Al interconnect on the chip side has a cross-section of 0.5 μm × 80 μm and a length of 300 μm. When we apply a current density of 10^5 A/cm² through the Al interconnect. What is Joule heating? When the current flows through the solder joint, which has a cylindrical cross-section of 100 μm in diameter and a height of 100 μm, what will be the Joule heating in the solder joint if we assume a uniform current density in the solder joint? Assume the resistivity of Al and solder to be 10^{-6} ohm-m and 10^{-5} ohm-cm, respectively.

8.4. What is the current density of an extension cord used at home for a 100-watt table lamp? Assume the Cu wire in the cord has a diameter of 0.1 mm and is 10 ft long, and the applied voltage is 110 volts.

8.5. Calculate the electric force of eE at a current density of 10^5 A/cm², and the chemical potential $\sigma\Omega$ at the elastic limit of Au, then calculate the critical length of Au.

REFERENCES

[1] P. Shewmon, *Diffusion in Solids* (2nd edn.), TMS, Warrendale, PA.
[2] V. S. Arpact, *Conduction Heat Transfer*, Addition-Wiley, Reading, MA (1966), p. 44.
[3] H. Ye, C. Basaran, and D. C. Hopkins, "Thermomigration in Pb-Sn solder joints under joule heating during electric current stressing," *Appl. Phys. Lett.*, 82, 1045–1047 (2003).
[4] Y. C. Chuang and C. Y. Liu, "Thermomigration in eutectic SnPb alloy", *Appl. Phys. Lett.*, 88, 174105 (2006).
[5] A. Huang, A. M. Gusak, K. N. Tu, and Y. S. Lai, "Thermomigration in SnPb composite flip chip solder joints," *Appl. Phys. Lett.*, 88, 141911 (2006).
[6] M. Li, D. W. Kim, S. Gu, D. Y. Parkinson, H. Barnard, and K. N. Tu, "Joule heating induced thermomigration failure in un-powered microbumps due to thermal crosstalk in 2.5D IC technology," *J. Appl. Phys.*, 120, 075105 (2016).
[7] K. N. Chen and K. N. Tu, "Materials challenges in three-dimensional integrated circuits," *MRS Bull.*, 40, 219–222 (2015).

CHAPTER 9

Effect of Oxidation on Electromigration

9.1 INTRODUCTION

In electromigration studies presented in the previous chapters, we have not considered oxidation. This is because in chip technology, all the Cu interconnects are protected by SiO_2, therefore no oxygen or moisture can reach them. However, in 3D IC packaging technology, the Cu interconnects, such as the redistribution layers (RDL), are exposed to ambient, and even though they may be covered by a layer of polymer, oxygen and moisture can reach them. Especially because electromigration on Cu occurs on the surface, so does oxidation; therefore, the effect of oxidation on electromigration or their interaction can become very serious, and the oxidation rate can be greatly accelerated [1–4].

No doubt, oxidation of metals has been a very difficult topic. This is because there are three kinds of diffusing species: cations, anions, and electrical charges. Their driving forces as well as kinetic processes are complicated. Now, we have to add electromigration to this difficult topic! We shall present experimental observations first.

9.2 EXPERIMENTAL

Figure 9.1 shows the test sample of RDL for electromigration (EM) in ambient. The cross-section of the sample has a width of 2 μm and a thickness of 3 μm. The length is 800 μm. The temperature during the EM test was 160°C. After the EM test, the longitudinal cross-section is shown in Figure 9.2, where a thick layer of Cu oxide has been found to form on top of the RDL. Also, the cross-sectional view is shown in Figure 9.3. In Figures 9.2 and 9.3, many interfacial voids are shown between the oxide layer and the remaining Cu layer, produced by electromigration. It is worth noting that the voids are formed in the Cu layer but not in the Cu oxide layer. This is because there

DOI: 10.1201/9781003384281-9

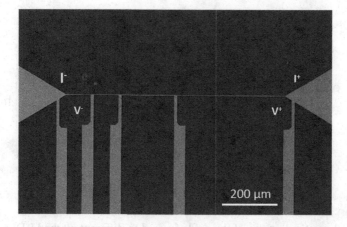

FIGURE 9.1   Plan-view SEM image of the Cu RDLs [5].

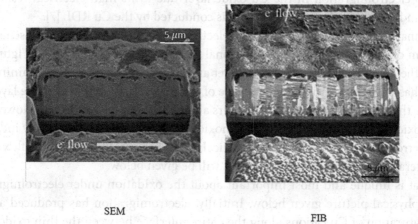

SEM   FIB

FIGURE 9.2   The longitudinal cross-section of test sample after EM test.

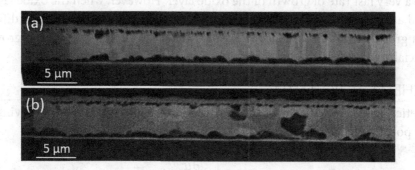

FIGURE 9.3   Failure analysis by cross-sectional SEM image along the long-axis of regular Cu when the line resistance increased 20% of the initial value: (a) at the cathode end and (b) at the anode end.

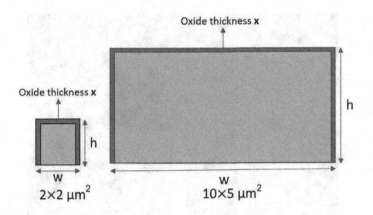

FIGURE 9.4 Schematic diagram of the oxide formed in different pitched RDLs.

was no electromigration in the Cu oxide layer due to its high electrical resistance, which indicates that the current density is conducted by the Cu RDL [7].

To analyse the interaction between electromigration and oxidation, a schematic diagram of the longitudinal cross-sectional view of the RDL is shown in Figure 9.4, where the upper layer is the Cu oxide layer and the bottom layer is the remaining Cu, which has not been oxidized yet. Because of the high resistivity of the oxide layer, we assume that electromigration of Cu occurs along the interface between the lower part of Cu oxide and the upper part of the unoxidized Cu. The horizontal arrow indicates the electromigration flux of J_{em}. The vertical arrow indicates the oxidation flux of J_{ox}. The interaction between these two fluxes will be given below.

What is unique and most important about the oxidation under electromigration is the physical picture given below. Initially, electromigration has produced a high supersaturation of Cu cations along the entire interface between the thin oxide layer and the remaining Cu layer. The Cu cations in turn produced an extremely large concentration gradient of Cu cations across the oxide layer in the vertical direction. It will lead to a very fast rate of growth of the oxide layer. However, when the oxide layer has reached a critical thickness, it itself becomes the barrier layer to Cu cation diffusion, and its growth tends to slow down or stop. What will follow then is the growth of interfacial voids driven by electromigration along the interface [4–6].

9.3 CHEMICAL POTENTIAL

In kinetics, when we consider atomic diffusion in a solid, we define the driving force to be a potential gradient.

$$F = -\frac{\partial \mu}{\partial x}$$

where μ is the chemical potential (a quantity with unit of energy) of an atom in the diffusion field at constant temperature and constant pressure.

$$\mu = \left(\frac{\partial G}{\partial C}\right) = \frac{\Delta G}{\Delta C}$$

Then, in the equation above, if we take $\Delta C = 1$, meaning the change of one atom or one mole of atoms,

$$\mu = \Delta G = \frac{\partial G}{\partial C}$$

Thus, the physical meaning of μ is the Gibbs free energy change when we change the composition by adding or removing one atom or one mole of atoms. For an ideal solution, it has been found that $\mu = kT \ln C$.

In Chapter 8 on stress, we have shown that the stress potential is $\mu = \pm \sigma \Omega$. Knowing the potential and, in turn, knowing the force, the atomic diffusion flux can be given below.

$$J = C \langle v \rangle = CMF = C \frac{D}{kT} \left(-\frac{\partial \mu}{\partial x}\right)$$

Below, we shall define the chemical potential of oxidation.

9.4 ORDINARY OXIDATION OF THIN FILM CU

A systematic study of ordinary oxidation of Cu thin films of thickness 20–150 nm in the temperature range of 100–300°C has been performed without electromigration. The oxide phase evolution goes from Cu_2O to CuO upon thermal oxidation in the temperature range of 100°C–450°C. The real-time synchrotron X-ray diffraction measurements show that the formation of the oxide phase depends critically on temperature and oxygen partial pressure. The formation of the CuO phase only starts after complete oxidation of the Cu thin film to the Cu_2O film. It was found that the oxidation kinetics of Cu_2O formation follow the linear rate law, which is attributed to an interfacial reaction-controlled process rather than a diffusion-controlled process. The rate-limiting process is taken to be oxygen dissociation at the gas-solid interface. Furthermore, a dramatic decrease in the linear oxidation rate is observed at low oxygen partial pressures [1–3].

The key difference between the ordinary oxidation of thin film of Cu and the rapid oxidation of Cu thin film under electromigration will be discussed.

9.5 INTERACTION BETWEEN OXIDATION AND ELECTROMIGRATION

On the interaction between oxidation and EM, we consider two stages. In the first stage, the effect of oxidation is strong, and all copper atoms driven by EM have been converted to Cu oxide, as shown in Figure 9.5. In the second stage, the oxide is thick enough to become a diffusion barrier, so the oxide growth can be ignored, and we have only electromigration. Consequently, the void formation introduced by electromigration appeared along the interface, or on top of the Cu layer, as well as at the bottom of the Cu layer, owing to the fact that at temperatures around 100°C electromigration in Cu occurs by surface diffusion.

In the first stage, we have the fluxes of EM and oxidation, respectively, as given below:

$$J_{EM} = C \frac{D_s}{kT} Z^* e \rho j$$

$$J_{ox} = C \frac{D}{kT} \left(-\frac{dE}{dy} \right) \tag{9.1}$$

where D_s is the surface diffusivity of Cu, D is the lattice diffusivity in Cu oxide, and dE/dy is the electro-chemical potential gradient in Cu oxide formation.

dE/dy is defined as a chemical (in general, electro-chemical) potential gradient across the oxide layer, which means the difference of chemical potentials of Cu at the interfaces of growing oxide, divided by the oxide thickness.

The difference in electro-chemical potentials of Cu **without current** is proportional to the thermodynamic driving force of the reaction, $\Delta g = g(Cu_2O) - (2/3)g(Cu) - (1/3)g(O)$. It is about $0.82*10^{-19}$ J/atom per atom [8–10].

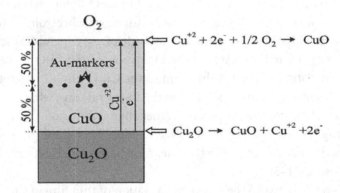

FIGURE 9.5 Schematic representation of the growth mechanism of CuO on the surface of Cu_2O.

Below, we consider the total number of Cu atoms transported by these two processes according to Eq. (9.1).

On electromigration, we have the number of atoms $= J_{em}At = J_{em}W\delta t$, where W is the width of the RDL line and δ is atomic thickness.

On oxidation, because it occurs on the entire surface of RDL, we have the number of atoms $= J_{ox}WLt$, where L is the total length of the RDL line and WL is the entire surface area of the RDL line.

In the first stage, it is critically important to define the link between EM and oxidation, which is given by the link between the number of atoms in EM as well as in oxidation. They are equal. This is because we assume that all Cu atoms (cations) in electromigration will be oxidized. Thus, we have

$$J_{em}W\delta t = J_{ox}WLt \quad \text{or}$$

$J_{em}\delta = J_{ox}L$ So, we obtain

$$J_{ox} = J_{em}\delta/L = C(D/kT)(-dE/dy) \tag{9.2}$$

which means that we can calculate the chemical potential of oxidation ($-dE/dy$) on the basis of Eq. (9.2).

Furthermore, the oxide growth rate can be calculated below, where Ω is the molar volume of Cu oxide.

$$J_{ox}WLt\Omega/WLt = J_{ox}\Omega \tag{9.3}$$

In the second stage, we ignore oxidation and only have EM, which occurs along the interface between the oxide layer and Cu. On the bottom surface of RDL, it has no oxidation, but if EM can occur along the bottom surface, the electromigration damage is typical.

REFERENCES

[1] N. Cabrera and N. F. Mott, "Theory of the oxidation of metals," *Rep. Prog. Phys.*, 12, 163–184 (1949).
[2] A. T. Fromhold, *Theory of Metal Oxidation*, North Holland Publishing Company, Amsterdam (1976).
[3] D. R. Gaskell, "Introduction to the thermodynamics of materials," in *Electrochemistry* (3rd edn.), Taylor & Francis, New York (1995) (Chapter 14).
[4] C.-F. Tseng, C.-S. Liu, C.-H. Wu, and D. Yu, "InFO (Wafer Level Integrated Fan-Out) Technology," 66th Electronic Comp. Technol. Conf., Las Vegas, NV, pp. 1–6, June 2016.
[5] I.-H. Tseng, P.-N. Hsu, T.-L. Lu, K. N. Tu, and C. Chen, "Electromigration failure mechanisms of <111>-oriented nanotwinned Cu redistribution lines with polyimide capping," *Results Phy.*, 24, 104154 (2021).

[6] I.-H. Tseng, P.-N. Hsu, W.-Y. Hsu, D.-P. Tran, B. T.-H. Lin, C.-C. Chang, K. N. Tu, and C. Chen, "Effect of oxidation on electromigration in 2-μm Cu redistribution lines capped with polyimide," *Results Phys.*, 31, 105048 (2021).

[7] Y. Liu, M. Li, D. W. Kim, S. Gu, and K. N. Tu, "Synergistic effect of electromigration and Joule heating on system level weak-link failure in 2.5D integrated circuits," *J. Appl. Phys.*, 118, 1–6 (2015). https://doi.org/10.1063/1.4932598.

[8] S. Choudhary, J. V. N. Sarma, S. Pande, S. Ababou-Girard, P. Turban, B. Lepine, and S. Gangopadhyay, "Oxidation mechanism of thin Cu films: a gateway towards the formation of single oxide phase," *AIP Adv.*, 8, 055114 (2018). https://doi.org/10.1063/1.5028407.

[9] Z. Grzesik and M. Migdalska, "Oxidation mechanism of Cu_2O and defect structure of CuO at high temperatures," *High Temperature Mater. Proc.*, 30, 277–287 (2011).

[10] E. Unutulmazsoy, C. Cancellieri, M. Chiodi, S. Siol, L. Lin, and L. P. H. Jeurgens, "In-situ oxidation studies of Cu thin films: growth kinetics and oxide phase evolution," *J. Appl. Phys.*, 127, 065101 (2020).

CHAPTER 10

Modified Mean-Time-to-Failure Equations

10.1 INTRODUCTION

To launch a new consumer electronic product, in addition to burn-in, the microelectronic industry requires product assurance based on statistical reliability tests. Especially, reliability against electromigration must be assured because it is the most common and persistent failure. In previous chapters, we have discussed the failure modes and understanding of the failure mechanism of electromigration. However, statistical analysis of failure distribution is needed in order to predict and to project the lifetime of a product. It can provide two pieces of very important information about the reliability issues of a product, to be given below.

Reliability has often been defined as "how quality changes over time." The difference between quality and reliability is that quality shows how well a device performs its proper function, while reliability shows how well this device can maintain its performance with time in field use.

To assure reliability, industry will first conduct a burn-in test to remove any weak link in the device so that no early failure can occur; thereafter, the device will fail according to its regular or normal lifetime, which is called the mean-time-to-failure (MTTF). For example, if we know the average life age of a person in Hong Kong is 80 years, if he/she dies at 60, he/she is an early failure, but if he/she lives to be 80 years old, he/she has a normal lifetime or is a normal failure. We shall discuss device burn-in in the next section. While burn-in has provided the critical function of removing early failures, it has no prediction capability to project the regular lifetime of a device in field use. For this purpose, we need to obtain the MTTF equation of a device.

About the MTTF equation, first, it is the equation to express the lifetime of a product about its use in the field of consumers. For example, we need to know the MTTF

DOI: 10.1201/9781003384281-10

of a device to be used under the hood of an automobile, which is a very hush environment with high temperatures.

Second, it is for the projection of future applications of an existing device, wherein the applied current density may have to be increased due to the increase in functionality.

Third, the electronic industry needs to know what the maximum current density, or I_{max}, is that can be applied to an existing device so that it can still perform well within a required lifetime without failure. In other words, the electronic industry wants to know what I_{max} is, which can be used in a given product. However, what is the standard method to calculate I_{max} that has not been defined in the literature? We shall provide a procedure to do so at the end of this chapter, in Section 10.8.

To conduct a statistical distribution analysis, there are two important requirements. First, we must have equipment that can measure the failure of a large number of devices as a function of time, temperature, and a specific driving force such as the applied current density in electromigration, which can stress the device until it fails. Second, we must have a large set of reproducible test samples, which should be as close as possible to the real consumer electronic products. Therefore, it is important to use samples provided by an industry laboratory, not arbitrary samples prepared in a campus environment.

In Figure 10.1, a set of equipment for statistical analysis of failure of the flip-chip solder joints under electromigration is shown. It consists of two furnaces, two power sources (for safety reasons, the applied current is limited to about 1 Amp), a multichannel control unit, and a personal computer for recording. They have been used to obtain the distribution of failure tested under two temperatures and two current densities, 2T2j. No doubt, it is better to have three furnaces and three power sources for 3T3j measurements.

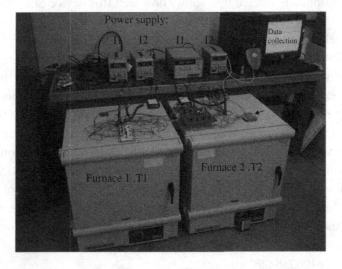

FIGURE 10.1 A set of equipment for statistical analysis of failure of flip-chip solder joints under electromigration.

On test samples, we take the above case of flip-chip solder joints. We can have multiple flip chips on a board for accelerated tests; therefore, we must be able to connect electrically several single flip chips together for the test so that a meaningful amount of data for statistical analysis can be obtained in a reasonable time.

Figure 10.2 shows an optical image of such a test board having four chips on it for electromigration tests. The layout of the solder joints between the chip and the board on one of the four chips is shown in Figure 10.3. The size of the chip is 0.3 cm × 0.3 cm, and the chip has 36 solder joints. The diameter of the solder joint is 250 μm. The contact opening at under-bump-metallization (UBM) is 200 μm in diameter. The chips are mounted on a printed circuit board using a Pb-free solder bump composed of Sn-1.2% Ag-0.5% Cu. The thin UBM is a layered thin film of Al/Ni(V)/Cu, where the thickness of Al, Ni(V), and Cu is 1 μm, 0.3 μm, and 1 μm, respectively. It is worth noting that the thickness of UBM has a strong effect on the failure mode or failure mechanism, as discussed in Chapter 8. A temperature sensor made of Pt serpentine wire is shown on the top surface of the Si chip for temperature calibration. The data from the tests will be presented and analysed in Section 10.2.1.

10.1.1 Burn-in

Burn-in is the process by which components of a system are examined before being used in service (often before the system is completely assembled from the components). This testing process can cause certain failures to occur under pre-determined conditions so that an understanding of the loading capacity of the component can be established.

The bathtub curve of burn-in, as shown in Figure 10.4, is widely used in reliability engineering and deterioration modelling. It describes a particular kind of curve of the failure function, which is composed of three parts. The first part has a decreasing

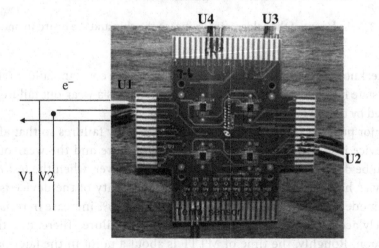

FIGURE 10.2 An optical image of such a test board having four chips on the board for electromigration tests.

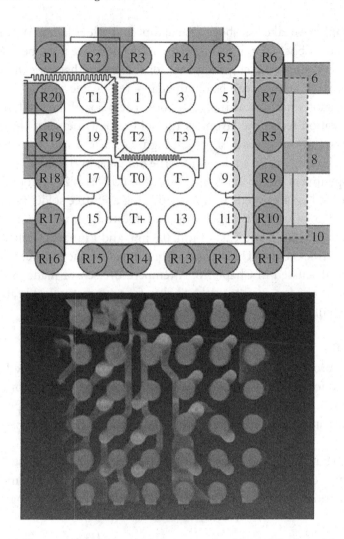

FIGURE 10.3 The layout of the solder joints between the chip and the board in one of the four chips.

failure rate, known as early failure. The second part has a constant failure rate, known as steady-state failure or regular failure. The third part is a wear-out failure. They are represented by dotted curves, respectively.

The major purpose of burn-in is to remove the early failures so that after burn-in, the device in use will follow the steady-state failure and the wear-out failure, which can be described by the MTTF equation. However, when the test resistance of the device has increased by about 20%, the reliability of the device is entering wear-out mode, and the device will fail soon. Thus, a 20% increase in resistance has been widely accepted as the beginning of wear-out failure. Thereafter, the device will fail soon. Roughly, the time of MTTF is about a point in the later part of the steady-state failure curve.

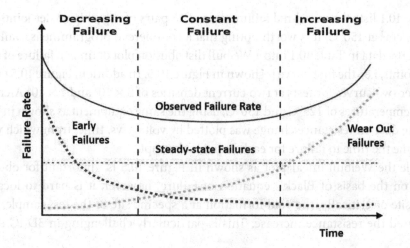

FIGURE 10.4   Typical bathtub-type burn-in curve.

10.2 STATISTICAL ANALYSIS OF ELECTROMIGRATION FAILURE

Before we discuss MTTF in electromigration, we shall revisit Black's equation in order to justify the dependence on the current density of j^{-2} in the equation. We shall present a unified model of MTTF, on the basis of entropy production that can be applied to the failure analysis of electromigration, thermomigration, and stress migration, as well as any other irreversible processes that may cause failure.

10.2.1 Black's Equation of MTTF for Electromigration

Black has provided an MTTF equation for electromigration as follows [1,2]:

$$\text{MTTF} = A\left(j^{-n}\right)\exp\left(\frac{E_a}{kT}\right) \tag{10.1}$$

where the time (MTTF) is related to temperature, T, and current density, j, by using three parameters: the pre-factor A, the current density power factor n, and the activation energy E_a.

It is of wide interest to point out that n is close to 2 in most experimental data of MTTF for electromigration, and E_a is related to the activation energy of atomic diffusion in electromigration. Why $n = 2$ has been a controversial topic for a long time, and it has been widely discussed [3–8]. In the next section, we shall derive a modified MTTF equation based on entropy production to show that $n = 2$ can be justified theoretically.

To determine these three parameters from experiments, we need to conduct electromigration tests at least under two temperatures and two current densities (2T2j) as a function of time until failure occurs. No doubt, it is better to do 3T3j.

Table 10.1 lists the measured failure time of 16 pairs of flip-chip solder joints tested at 10^4 A/cm² at 150°C. It is worth noting that a commercial programme is available to convert the data in Table 10.1 into a Weibull distribution plot of time to failure of a set of solder joints, i.e., the first curve is shown in Figure 10.5. In addition, Figure 10.5 exhibits the other two curves for tests at two current densities of 5×10^3 and 1×10^4 A/cm² and at two temperatures of 125°C and 150°C, using the same equipment as shown in Figure 10.1. The data on resistance change was plotted by voltage vs. time, from which we can determine the time to failure for each individual sample.

While the Weibull linear plot as shown in Figure 10.5 is important for obtaining MTTF on the basis of Black's equation of failure, however, it is hard to locate the failure site or critically, the void formation at a specific site of the test sample, which has caused the resistance increase. This is particularly challenging in 3D IC devices

TABLE 10.1   Time-to-Failure (TTF) of 16 Pairs of Flip-Chip Solder Bumps Tested at 10^4 A/cm² and 150°C

Sample	TTF	Sample	TTF
8_T1_I1_U1	47	8_T1_I3_U1	59.5
8_T1_I1_U2	53	8_T1_I3_U2	68.5
8_T1_I1_U3	55.5	8_T1_I3_U3	43.5
8_T1_I1_U4	103	8_T1_I3_U4	58
8_T1_I2_U1	47.5	8_T1_I4_U1	68.5
8_T1_I2_U2	81.5	8_T1_I4_U2	25
8_T1_I2_U3	45.5	8_T1_I4_U3	41
8_T1_I2_U4	48	8_T1_I4_U4	53.5

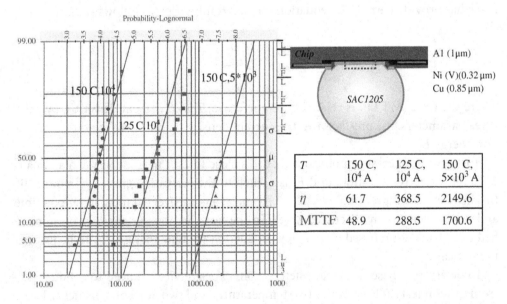

FIGURE 10.5   The Weibull distribution plots of time to failure of a set of solder joint samples.

because it is not easy to locate the failure site in a 3D structure, which has led to an open circuit. To do so, a high-resolution in-situ synchrotron radiation X-ray tomographic inspection is needed.

Often, it is asked why we tend to use the Weibull distribution in electromigration studies rather than other kinds of distribution functions. This is because it has been found that in the Weibull distribution, we can observe the data of early failure easily due to weak-link failures in the test samples. From the viewpoint of the system reliability of a device, early failure is the most undesirable. They must be removed by burn-in. Another reason is the correlation to the physical mode of time-dependent failure in electromigration, where the exact similarity between the Weibull distribution function and the Johnson-Mehl-Avrami (JMA) equation of phase transformations is intriguing, which will be discussed later.

It is worth pointing out that if a failure occurs following the Weibull distribution, it is a normal failure, so it is acceptable. Only the early failure is unacceptable, and it must be removed.

To find the activation energy in Black's MTTF equation, we use the data at the same current density with two temperatures and the relation below:

$$\frac{MTTF_2}{MTTF_1} = \frac{\exp\left(\dfrac{E_a}{kT_2}\right)}{\exp\left(\dfrac{E_a}{kT_1}\right)}$$

$$E_a = \left(\frac{1}{kT_2} - \frac{1}{kT_1}\right)\log\left(\frac{MTTF_2}{MTTF_1}\right) \tag{10.2}$$

Then, to determine "n," we can use the data at the same temperature with two different current densities and the relation below:

$$\frac{MTTF_2}{MTTF_1} = \left(\frac{j_1}{j_2}\right)^n \tag{10.3}$$

However, due to Joule heating, we have to be very careful in applying the above two relations. When we calculate the activation energy using Eq. (10.2), we have to take into account the effect of Joule heating on temperature by changing kT to $k(T + \Delta T)$, where ΔT is due to Joule heating at the applied current density and it has been measured, so kT_1 becomes $k(T_1 + \Delta T)$ and kT_2 becomes $k(T_2 + \Delta T)$ in Eq. (10.2). This correction is rather straight-forward.

When we calculate "n" using Eq. (10.3), it is harder. Because Joule heating is different at different current densities, the temperature cannot be assumed to be the same for two different current densities measured at the same annealing temperature. The actual temperature is not the same, so the problem is non-trivial. Therefore, we need to know the ΔT increase at different current densities in order to correct the temperature. By using a temperature sensor with a Pt thin film serpentine line deposited on the test sample, we can obtain the temperature calibration in order to determine the specific temperature increase.

The activation energy was obtained to be 1.15 eV/atom, on the basis of the data shown in Figure 10.5. Also, we found $n = 2.08$. Then we can determine the pre-factor, $A = 1.5 \times 10^{-1}$ sec-amp^2/cm^4. On the basis of the obtained MTTF equation, the prediction or extrapolation of the life time of a device at the condition of field use by consumers can be calculated.

10.2.2 Weibull Distribution Function and JMA Theory of Phase Transformations

The Weibull distribution function is given as

$$F(t) = 1 - \exp\left[-\left(\frac{t}{\eta}\right)^{\beta}\right] \tag{10.4}$$

where $F(t)$ is the fraction of the failed sample as a function of time, η is the characteristic lifetime, and β is the shape factor or the slope of the Weibull plot, where a larger slope will give a narrower distribution of time to failure.

There is a strong similarity in the mathematical form between the Weibull distribution function and the JMA canonical equation of phase transformations [9]. In classical phase transformations, for example, the crystallization of an amorphous phase to a crystalline phase without changing composition, the fraction of the transformed volume is expressed by the JMA equation as

$$X_T = 1 - \exp(-X_{ext}) = 1 - \exp\left[-\left(\frac{t}{\lambda}\right)^{n}\right] \tag{10.5}$$

where X_T is the fraction of the volume transformed and X_{ext} is the fraction of the extended volume, which is defined as

$$X_{ext} = \int_{\tau=0}^{\tau=t} \frac{4\pi}{3} R_N R_G^3 (t-\tau)^3 d\tau \tag{10.6}$$

where R_N is the nucleation rate, R_G is the growth rate of transformed spherical particles, and t is the time of transformation. In essence, the physical meaning of X_{ext} is the sum of the volume of all spherical particles in the period of $\tau = 0$ to t, without considering growth impingement and phantom nucleation.

Because the mathematical form of the Weibull distribution function, Eq. (10.4), and the JMA equation of phase transformations, Eq. (10.5), are similar, it is of interest to correlate them so that we may obtain a physical explanation of the Weibull distribution of MTTF from the point of view of the phase transformation of void formation in a set of interconnects. It means we may regard electromigration-induced failure of an open circuit due to void formation as a phase transformation in the cathode end of an Al or Cu line or in a flip-chip solder joint where a void nucleates and grows. No doubt, this is a topic of interdisciplinary that requires careful study in the future.

In Black's equation of MTTF, the physical meaning of the activation energy has not been defined clearly. It could mean the activation energy of nucleation of the void, or the growth of the void, or the sum of both, as indicated by R_N and R_G in Eq. (10.6). However, if the failure time is short, which indicates that the tests are conducted at a high temperature with a high current density, the nucleation and growth of voids are fast. On the other hand, if the failure time is long, it indicates that the tests are conducted at a low temperature with a low current density. No doubt, the latter could have a better correlation to the actual field use of the product, but it will take a greater effort and longer time to finish the tests. At the end of this chapter, we will have a deeper discussion on how to overcome this issue.

10.3 A UNIFIED MODEL OF MTTF FOR ELECTROMIGRATION, THERMOMIGRATION, AND STRESS MIGRATION BASED ON ENTROPY PRODUCTION

10.3.1 Revisit Black's Equation of MTTF for Electromigration

First, we take J_e to be the atomic flux driven by electromigration, and we have

$$J_e = C \frac{D}{kT} Z^* e \rho j \qquad (10.7)$$

On the driving force of electromigration, we have

$$X_e = Z^* eE = Z^* e\rho j, \qquad (10.8)$$

where $E = -d\phi/dx = \rho j$ is the electric field.

Now, let V^* be the critical volume of the void formation at the cathode end, which has led to an open failure, and we have $V^* = \Omega JAt$, where Ω is the atomic volume, A is the cross-section of the diffusion, and t (or MTTF) is the time to failure, so

$$t = V^* / \Omega AJ. \qquad (10.9)$$

The above equation shows that MTTF is proportional to $1/j$, thus $n = 1$. But, in Black's equation, why is n taken to be 2?

To revisit Black's equation, we now consider entropy production in irreversible processes as the basis of microstructure failure, as presented in Chapter 4. According to Onsager, the entropy production rate is equal to

$$\frac{TdS}{Vdt} = JX \tag{10.10}$$

where T is temperature, dS/dt is the rate of entropy production, V is the volume of the sample in the irreversible process, and J and X are the conjugated flux and conjugated driving force, respectively.

From Eq. (10.10), the total entropy production is, $S = (dS/dt)t^{failure}$, until failure is $TS/V = J_e X_e t^{failure}$. We have

$$t^{failure} = MTTF = \frac{TS_{threshold}}{VJ_e X_e} = \frac{TS_{threshold}}{V} \frac{1}{\left(C\dfrac{D}{kT}Z^*e\rho j\right)(Z^*e\rho j)}$$

$$= \frac{TS_{threshold}}{V} \frac{1}{C} \frac{kT}{D} \frac{1}{(Z^*e\rho)^2} j^{-2} = \frac{S_{threshold}}{VC} \frac{kT^2}{(Z^*e\rho)^2 D_0} \exp\left(\frac{E}{kT}\right) j^{-2} \tag{10.11}$$

$$= A\left(\frac{T}{j}\right)^2 \exp\left(\frac{E}{kT}\right)$$

where $A = \dfrac{S_{threshold}kT^2}{VCD_0(Z*e\rho)^2}$. The above equation shows that $n = 2$ is justified theoretically in Black's equation. Yet, it is $(T/j)^2$, not $(1/j)^2$.

10.3.2 Joule Heating Has No Effect on Electromigration Damage

In the above derivation, entropy production in electromigration is used to estimate the failure time, or MTTF. But we should make it clear that the Joule heating of electrical conduction does not contribute to electromigration, except for the temperature increase. We have already discussed this point in Section 8.3.1 of Chapter 8. Also, please see Figure 3.3 and Eq. (3.2) in Chapter 3.

10.4 NEW MTTF EQUATION FOR THERMOMIGRATION

To obtain MTTF for thermomigration, the conjugated atomic flux is given as follows [10]:

$$J_h = C\frac{D}{kT}\frac{Q^*}{T}\left(-\frac{\partial T}{\partial x}\right) \tag{10.12}$$

where Q^* is defined as the heat of transport in thermomigration, and Q^* has the same dimension as μ, so it is the heat energy per atom. The conjugated driving force is

$$X_h = \frac{Q^*}{T}\left(-\frac{dT}{dx}\right)$$

(10.13)

We have $J_h X_h t^{failure} = TS_{threshold}/V$, so

$$\text{MTTF} \approx t^{failure} = \frac{TS_{threshold}}{V J_h X_h} = B\left(-\frac{dT}{dx}\right)^{-2}\exp\left(\frac{E_a}{kT}\right)$$

(10.14)

10.5 NEW MTTF EQUATION FOR STRESS MIGRATION

Stress potential is defined as σΩ, where σ is stress and Ω is atomic volume. Thus, the conjugated driving force of stress migration is given as follows [10]:

$$X_S = -\frac{d\sigma\Omega}{dx}$$

(10.15)

And the conjugated atomic flux in stress migration is given as

$$J_S = CMF = C\frac{D}{kT}\left(-\frac{d\sigma\Omega}{dx}\right) = \frac{D}{kT}\left(-\frac{d\sigma}{dx}\right)$$

(10.16)

We have $J_S X_S t^{failure} = TS_{threshold}/V$, so

$$\text{MTTF} \approx t^{failure} = \frac{TS_{threshold}}{V J_S X_S} = G\left(-\frac{d\sigma}{dx}\right)^{-2}\exp\left(\frac{E_a}{kT}\right)$$

(10.17)

10.6 THE LINK AMONG MTTF EQUATIONS FOR ELECTROMIGRATION, THERMOMIGRATION, AND STRESS MIGRATION

The conjugated forces of electromigration, thermomigration, and stress migration in Sn or Sn-rich solder have been shown to be nearly the same in Chapter 6. This provides the justification for the following analysis of the MTTF of three of them by equating their driving forces.

$$Z^*e\rho j = 3k\frac{dT}{dx} = -\frac{d\sigma\Omega}{dx}$$

(10.18)

Thus, we have

$$j = \frac{3k}{Z^*e\rho}\frac{dT}{dx} = \frac{-1}{Z^*e\rho}\frac{d\sigma\Omega}{dx}$$

(10.19)

Then, by substituting j^{-2} into the modified Black's equation, we can obtain MTTF for thermomigration and stress migration, respectively, below:

$$\text{MTTF}_h = A \left(\frac{3k}{Z^* e \rho} \frac{dT}{dx} \right)^{-2} \exp\left(\frac{E_a}{kT} \right) \tag{10.20}$$

$$\text{MTTF}_S = A \left(\frac{1}{Z^* e \rho} \frac{d\sigma\Omega}{dx} \right)^{-2} \exp\left(\frac{E_a}{kT} \right) \tag{10.21}$$

10.7 STUDY OF MTTF WITH 1T1J

Onsager's equation of entropy production shows that production depends on driving force and flux in an irreversible process. If we can lower down the driving force or slower down the flux, we will have low entropy production, which is significant.

When $n = 2$ is fixed in the MTTF equation of electromigration, the remaining variables are the activation energy and the pre-factor. For interconnect materials such as Al and Cu, the activation energy of atomic diffusion is known. Furthermore, in the case of Pb-free solder joints, the activation energy is about 1 eV. Therefore, the only remaining unknown variable in the MTTF equation is the pre-factor. It means we can greatly simplify the effort to determine MTTF for future devices. Specifically, we need only to do 1T1j, and there is no need to do 3T3j experiments, meaning that indeed we can save a lot of effort.

Then, we ask how to choose 1T1j. Typically, it should have a low temperature and a low applied current density, and they should be as close to the user condition in field use as possible. No doubt, it cannot be too close; otherwise, we have no acceleration effect in the test.

On the other hand, if we choose a high temperature and a high current density, we could obtain data faster. However, we might introduce a failure mechanism that is not the same as that in field use, so we could obtain a wrong conclusion on MTTF.

10.8 HOW TO DETERMINE I_{MAX}?

If a device has performed very well in applications, the electronic company that produces the device would like to consider if it could become a next-generation device with wider applications. Typically, this means the applied current density must be increased due to the addition of more functions. Then, the question becomes: how large can the current density be increased in a given device without affecting the MTTF? In other words, what is I_{max}?

We shall consider the performance of a device below. When the user condition is at a temperature of 100°C and the applied current density is 2×10^3 A/cm², we have calculated its MTTF1 = 13.2 years.

Now, to calculate I_{max} for projection, we assume that the MTTF2 in future use is reduced to 5 years, which is the typical value. If we keep all the other parameters the same, the current density change is given by

$$\frac{\text{MTTF1}}{\text{MTTF2}} = \frac{13.2}{5} = \left(\frac{j_1}{j_2}\right)^{-2}$$

$$j_2 = \left[\left(\frac{\text{MTTF1}}{\text{MTTF2}}\right)(j_1)^2\right]^{\frac{1}{2}}$$

$$j_2 = 3.24 \times 10^3 \, \text{A/cm}^2$$

Thus, we have $I_{max} = 3.24 \times 10^3 \, \text{A/cm}^2$.

PROBLEMS

10.1. What is mean-time-to-failure? Why is it important?

10.2. Electromigration has been a major reliability concern for microelectronic devices. How come we have almost never found that our personal computer, laptop computer, and cell phone have failed due to electromigration?

10.3. In Black's equation, we have measured the activation energy to be $E = 1$ eV/atom, the power factor of current density is $n = 2$, and the pre-factor $A = 10^{-2}$ sec $(\text{A/cm}^2)^2$. Calculate the MTTF when the device is used at 100°C with a current density of 2×10^3 A/cm². Then, for the next-generation device, when the current density is increased to 1×10^4 A/cm², what will be the MTTF?

10.4. What is the difference between yield and reliability problems in microelectronic technology? Please use examples to explain them.

10.5. In Black's equation of MTTF, why is the sign in front of the activation energy positive?

REFERENCES

[1] I. Prigogine, *Introduction to Thermodynamics of Irreversible Processes* (3rd ed.), Wiley-Interscience, New York (1967).

[2] K. N. Tu and A. M. Gusak, A unified model of mean-time-to-failure for electromigration, thermomigration, and stress-migration based on entropy production, *J. Appl. Phys.*, 126, 075109 (2019). https://doi.org/10.1063/1.511115.

[3] K. N. Tu, *Electronic Thin-Film Reliability*, Cambridge University Press, Cambridge (2010).

[4] M. Ding, G. Wang, B. Chao, P. S. Ho, P. Su, and T. Uehling, "Effect of contact metalliza- tion on electromigration reliability of Pb-free solder joints," *J. Appl. Phys.*, 99(9), 094906 (2006).

[5] H.-Y. Chen and C. Chen, "Measurement of electromigration activation energy in eutec- tic SnPb and SnAg flip-chip solder joints with Cu and Ni under-bump metallization," *J. Mater. Res.*, 25(9), 1847–1853 (2010).

[6] C.-K. Hu, R. Rosenberg, and K. Y. Lee, "Electromigration path in Cu thin-film lines," *Appl. Phys. Lett.*, 74(20), 2945–2947 (1999).

[7] P. Su, L. Li, Y. S. Lai, T. Y. Chiu, and C. L. Kao, "A Comparison Study of Electromigration Performance of Pb-free Flip Chip Solder Bumps," 2009 59th Electronic Components and Technology Conference, IEEE, p. 903, 2009. https://doi.org/10.1109/ECTC.2009.5074.

[8] J.J. Clement, and C.V. Thompson,"Modeling electromigration induced stress evolution in confined metal lines," *Journal of Applied Physics* 78, 900–904 (1995).

[9] D.A. Porter, and K.E., "Eastering, Phase Transformations in Metals and Alloys," 2e, 290 (1992). London: Chapman & Hall.

[10] K.N. Tu, and A.M. Gusak, "A unified model of mean-time-to-failure for electromigra- tion, thermomigration, and stress-migration based on entropy production", *Journal of Applied Physics* 126, 075109 (2019).

Index

Printed in the United States
by Baker & Taylor Publisher Services